Organic Electronics - From Fundamentals to Applications

Edited by Arun Dev Dhar Dwivedi,
Debashish Maji
and Parthsarathi Chakrabarti

Published in London, United Kingdom

Organic Electronics - From Fundamentals to Applications
http://dx.doi.org/10.5772/intechopen.111086
Edited by Arun Dev Dhar Dwivedi, Debashish Maji and Parthsarathi Chakrabarti

Contributors
Aaron DiFilippo, Amrita Chakraborty, Arun Dev Dhar Dwivedi, Debashis Maji, Elham Bigdeli, Jurij Urbančič, Kadiyam Anusha, Marius K. Orlowski, Parthasarathi Chakrabarti, Ram Babu Pareek, Reza Najjar, Shubham Dadhich

First published in London, United Kingdom, 2025 by IntechOpen
IntechOpen is the global imprint of INTECHOPEN LIMITED, registered in England and Wales, registration number: 11086078, 167-169 Great Portland Street, London, W1W 5PF, United Kingdom

For EU product safety concerns: IN TECH d.o.o., Prolaz Marije Krucifikse Kozulić 3, 51000 Rijeka, Croatia, info@intechopen.com or visit our website at intechopen.com.

British Library Cataloguing-in-Publication Data
A catalogue record for this book is available from the British Library

Organic Electronics - From Fundamentals to Applications
Edited by Arun Dev Dhar Dwivedi, Debashish Maji and Parthsarathi Chakrabarti
p. cm.
Print ISBN 978-1-83634-083-6
Online ISBN 978-1-83634-082-9
eBook (PDF) ISBN 978-1-83634-084-3

If disposing of this product, please recycle the paper responsibly.

Meet the editors

Arun Dev Dhar Dwivedi received his Ph.D. degree in Electronics Engineering from the Indian Institute of Technology (IIT) BHU Varanasi, India, in 2010. Currently, he is working as an Associate Professor (Senior) in the School of Electronics Engineering (SENSE), Department of Micro and Nano Electronics at Vellore Institute of Technology (VIT), Vellore, Tamil Nadu, India. He has been awarded the Early Career Research Award (2017) from the Science and Engineering Research Board (SERB), Department of Science and Technology (DST), Government of India. He completed a SERB DST-funded project entitled "Numerical simulation and compact modeling of organic thin film transistors for future flexible electronics" from 2017 to 2020. He has published more than 55 papers in International Journals and 40 papers in conferences and seminars. He is a reviewer for various international journals, including *IEEE Transactions on Electron Devices, Organic Electronics, Journal of Electronic Materials,* and *Journal of Computational Electronics*, among others. His research interests include micro- and nanoelectronic devices and circuits, numerical simulation and compact modelling of OTFTs, metal oxide semiconductor-based thin film transistors, flexible electronics and optoelectronics, AI, and ML.

Debashis Maji completed his Ph.D. at the School of Medical Science and Technology, Indian Institute of Technology Kharagpur, in 2015. Currently, he has been working as an Associate Professor (Grade 1) in the Department of Sensors and Biomedical Technology at the School of Electronics Engineering, Vellore Institute of Technology, Vellore, since May 2017. He has almost eight years of experience in developing flexible sensors for biomedical applications. His expertise includes surface modification techniques for polymers, non-conventional fabrication strategies, design and development of wearable and conformal sensors, BioMEMS, Flexible biosensors, and microfluidic biochips for biomedical applications. He has completed one SERB-funded project (SERB-ECR) on the development of a flexible and wearable microelectrode array for accelerated wound healing applications. He is currently working on a wearable skin monitoring patch for point-of-care applications. He has filed three Indian patents and authored over 40 research papers in reputable national and international journals and conference proceedings, including those published by ACS, IOP, IEEE, and Springer Nature's Scientific Reports, across various multidisciplinary fields. He is an active member of IEEE and the IEEE Engineering in Medicine and Biology Society. He is an active reviewer for several reputable journals, including Nature Flexible Electronics, Scientific Reports, Sensors and Actuators, IEEE, and MDPI, among others. He has been an active member of the prestigious IEEE & IEEE EMB societies for more than 15 years.

Parthsarathi Chakrabarti, Former Director of the Indian Institute of Engineering Science and Technology, Shibpur (2018-2024), and NIT Allahabad (2011-2016), is a former Professor at IIT (BHU), Varanasi. He has made a significant contribution to the area of Microelectronics and Photonics. He completed several sponsored R&D projects and guided over twenty Ph.D. scholars. He has over 300 research papers and 4 granted patents to his credit. He is a recipient of the Visiting Fellowship of the Indian National Science Academy (INSA), the Senior Visiting Fellowship of the EPSRC, UK, and the MRSI Distinguished Lectureship Award 2019-20. Prof. Chakrabarti is a Fellow of the Institution of Engineers and a Senior Member of IEEE, USA.

Contents

Preface

The field of organic electronics has undergone transformative growth over the past two decades, evolving from a niche area of research into a cornerstone of next-generation electronic and optoelectronic technologies. This edited book, *Organic Electronics – From Fundamentals to Applications*, has been carefully curated to provide a comprehensive and coherent overview of this dynamic domain. It serves as a bridge between fundamental scientific principles and the diverse range of practical applications emerging from organic electronic materials and devices.

The inspiration for this book stems from the growing demand for flexible, lightweight, low-cost, and environmentally friendly alternatives to conventional inorganic electronics. Organic semiconductors, conductive polymers, and related materials now play a vital role in a broad spectrum of applications, including organic light-emitting diodes (OLEDs), organic photovoltaics (OPVs), organic field-effect transistors (OFETs), and bio-integrated electronics. Each chapter in this volume has been contributed by leading experts in their respective areas, offering readers both foundational knowledge and insights into the latest advancements.

This book is structured to cater to a wide readership. The initial chapters lay the groundwork by exploring the chemistry, physics, and materials science underpinning organic electronic devices. Subsequent chapters delve into device architecture, fabrication techniques, and performance optimization strategies. The final sections are dedicated to emerging applications and future perspectives, highlighting how organic electronics are enabling innovations in fields as varied as wearable technology, energy harvesting, and biomedical devices.

We hope this book will serve as a valuable resource for students, researchers, and professionals engaged in materials science, electrical engineering, applied physics, and related disciplines. By presenting a balanced treatment of theory and application, our goal is to foster further exploration and collaboration in this rapidly evolving interdisciplinary field.

We would like to extend our sincere gratitude to all contributors for their expertise and dedication, as well as to the reviewers for their constructive feedback. Special thanks are due to Kristina Kardum Cvitan, Publishing Process Manager at IntechOpen Limited, UK, for her unwavering support, guidance, and professionalism throughout the publishing process. Her contributions have been instrumental in bringing this book to fruition.

We also extend our heartfelt thanks to Prof. M. Jasmine Pemeena Priyadarsini, Dean of the School of Electronics Engineering, Vellore Institute of Technology (VIT), Vellore, and Prof. Sriadibhatla Sridevi, Head of the Department of Micro and Nano Electronics,

School of Electronics Engineering, VIT, Vellore, India. Their encouragement, leadership, and continuous support have been pivotal in shaping the academic environment that fostered the development of this book.

We trust that *Organic Electronics – From Fundamentals to Applications* will inform, inspire, and energize new discoveries and applications in the exciting world of organic electronics.

Dr. Arun Dev Dhar Dwivedi and Dr. Debashis Maji
School of Electronics Engineering,
Vellore Institute of Technology,
Vellore, India

Prof. Parthasarathi Chakrabarti
Professor (Retired),
Department of Electronics Engineering,
IIT (BHU) Varanasi, India

Section 1

Organic Electronics: Fundamentals

Chapter 1

Introductory Chapter: Organic Electronics – From Fundamentals to Applications

Arun Dev Dhar Dwivedi, Debashis Maji and Parthasarathi Chakrabarti

1. Introduction

Organic electronics is an emerging and interdisciplinary field that combines principles of organic chemistry, materials science, solid-state physics, and electrical engineering to develop electronic devices using carbon-based (organic) materials. Unlike traditional silicon-based electronics, organic electronics exploit the electrical properties of π-conjugated organic molecules and polymers, which offer the advantages of low-temperature processing, mechanical flexibility, tunable optical/electrical properties, and compatibility with large-area fabrication techniques such as inkjet printing and roll-to-roll processing [1–31].

The promise of lightweight, flexible, and cost-effective electronic systems has accelerated research and commercialization in areas such as organic light-emitting diodes (OLEDs), organic photovoltaics (OPVs), organic field-effect transistors (OFETs), and organic sensors [1–31]. As we move toward a future where electronics become increasingly integrated into daily life—embedded into clothing, packaging, and medical devices—organic electronics are positioned to play a vital role.

2. Historical perspective

The development of organic electronics traces back to the discovery of conductive polymers by Alan J. Heeger, Alan G. MacDiarmid, and Hideki Shirakawa in the 1970s, a breakthrough that led to their Nobel Prize in Chemistry in 2000 [1–31]. Subsequent advancements in material science and fabrication techniques have propelled organic electronics into commercial viability, particularly in display technologies and renewable energy solutions [1–31]. In subsequent years, the development of materials such as polythiophene, polyaniline, and poly(p-phenylene vinylene) (PPV) marked significant milestones in the field.

The 1980s and 1990s witnessed the birth of organic light-emitting diodes and organic thin-film transistors, transforming theoretical concepts into functional devices. With the commercial success of OLED displays in smartphones and televisions, the field has gained considerable industrial traction.

IntechOpen

3. Fundamentals of organic semiconductors

Organic semiconductors are composed of π-conjugated systems in which delocalized electrons allow for charge transport. These materials can be small molecules or polymers and can be synthesized to tailor electrical, optical, and mechanical properties. The key properties governing their behavior include:

- *Molecular orbital theory*: Charge transport in organic semiconductors is governed by the highest occupied molecular orbital (HOMO) and the lowest unoccupied molecular orbital (LUMO), analogous to the valence and conduction bands in inorganic semiconductors [2].

- *Charge carrier mobility*: Typically lower than in inorganic semiconductors, often in the range of 10^{-6} to 10 $cm^2/V{\cdot}s$. Mobility is influenced by molecular packing, crystallinity, and disorder.

- *Exciton formation*: Organic semiconductors exhibit tightly bound excitons due to their low dielectric constants, which must be dissociated for efficient charge generation in devices like OPVs [3].

Organic electronic devices rely on organic semiconductors, which exhibit semiconducting properties due to the conjugated π-electron systems in their molecular structure. These materials can be classified into:

- *Small molecules*: e.g., pentacene and rubrene, which are often used in evaporated thin films.

- *Polymers*: e.g., poly(3-hexylthiophene) (P3HT), which can be solution-processed for large-area applications (**Figure 1** and **Table 1**).

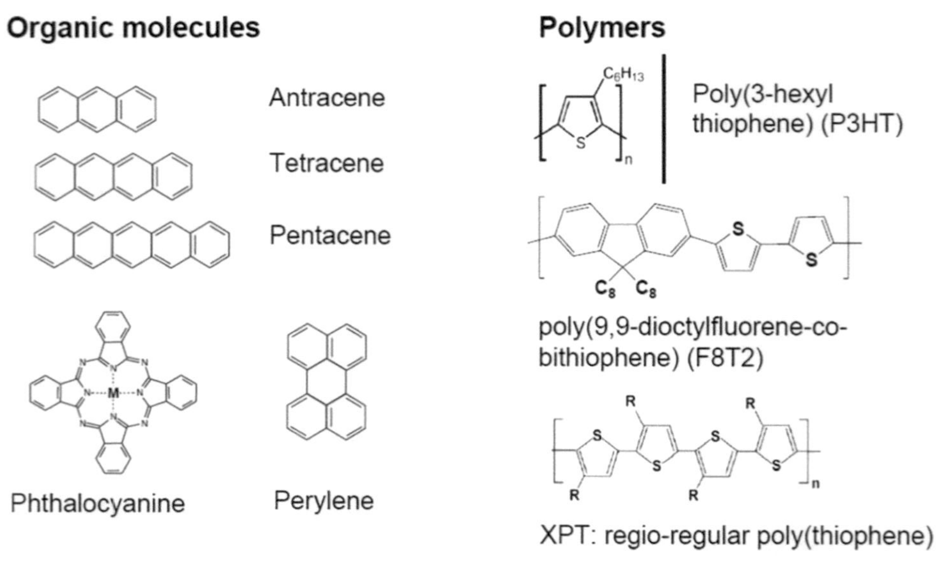

Figure 1.
Common organic semiconductor materials divided into organic molecules and polymers [2, 32, 33].

Advantages	Disadvantages
Low-cost processing	Low carrier mobility
Large-area compatible	Stability
Low-temperature processing	Patterning of films
Tailoring of electronic and optical properties	Novel fabrication technology required
Certain properties not easily attainable with conventional materials	

Table 1.
Advantages and disadvantages of organic and polymer semiconducting materials [2, 32].

3.1 Linear combination of atomic orbitals (LCAO) and molecular orbital theory (MOT)

Understanding the electronic structure of organic semiconductors begins at the molecular level, where individual atomic orbitals combine to form molecular orbitals. These molecular orbitals govern the optical, electrical, and charge transport properties of organic materials.

3.1.1 Linear combination of atomic orbitals (LCAO)

The LCAO method is a foundational approximation used in quantum chemistry to construct molecular orbitals by combining the wavefunctions of individual atomic orbitals. For two atoms A and B with atomic orbitals ψ_A and ψ_B, a molecular orbital ψ_M is given by:

$$\psi_M = c_A \psi_A + c_B \psi_B \tag{1}$$

where c_A and c_B are coefficients that determine the contribution of each atomic orbital. The constructive and destructive interference of atomic orbitals leads to the formation of *bonding* and *antibonding* molecular orbitals, respectively.

In organic π-conjugated systems, such as benzene or polyacetylene, the $2p_z$ orbitals of carbon atoms overlap laterally to form delocalized π-molecular orbitals that extend over the entire molecule. These delocalized orbitals are essential for charge transport in organic semiconductors.

3.1.2 Molecular orbital theory (MOT)

Molecular Orbital Theory extends the LCAO approach to describe the electronic structure of entire molecules, treating electrons as delocalized over the whole molecular framework rather than localized between specific atom pairs (as in valence bond theory).

3.1.2.1 Key features of MOT in organic electronics include

- *Highest Occupied Molecular Orbital (HOMO)* and *Lowest Unoccupied Molecular Orbital (LUMO)*: Analogous to the valence and conduction bands in inorganic semiconductors. The HOMO-LUMO gap determines the optical absorption and electronic bandgap of the material.

- *π and π** orbitals: In conjugated organic molecules, the π-bonding orbitals are filled, and the π*-antibonding orbitals are empty, allowing for electronic transitions upon photoexcitation.
- *Orbital delocalization*: Greater delocalization leads to improved charge transport by reducing reorganization energy and enabling electronic coupling between adjacent molecules.

The alignment of HOMO and LUMO levels with electrode work functions or with donor/acceptor materials is critical in designing efficient devices like organic field-effect transistors (OFETs), organic photovoltaics (OPVs), and organic light-emitting diodes (OLEDs).

3.1.3 Effect of bonding on properties of organic semiconductors

The *bonding in organic semiconductors*, particularly the nature of *π-conjugation*, has a profound impact on their *electronic, optical, and transport properties*. Here is a breakdown of how bonding affects these materials:

3.1.3.1 Delocalized π-bonding → electronic properties

In organic semiconductors, *alternating single and double bonds* (π-conjugation) allow p_z *orbitals* to overlap and form *delocalized molecular orbitals*.

3.1.3.1.1 Effects

- Forms *HOMO* (π) and *LUMO* (π*) orbitals.
- The *HOMO-LUMO gap* determines the *bandgap*.
    - *Smaller gap* → better conductivity.
    - *Tunable* via chemical structure (e.g., adding electron-donating or withdrawing groups).

3.1.3.2 Bonding and optical properties

The *extent of conjugation* affects light absorption and emission:

3.1.3.2.1 Effects

- *Longer conjugation* → lower HOMO-LUMO gap → *absorbs longer wavelengths* (red-shift).
- Useful for *OLEDs* and *organic photovoltaics* (OPVs), where absorption in the visible range is desired.
- Fluorescence and phosphorescence depend on π-π* transitions.

3.1.3.3 Bonding affects charge transport

Conjugated π-systems provide *pathways for electron delocalization*, which is essential for charge mobility.

3.1.3.3.1 Effects

- *Stronger π-π stacking* between molecules (intermolecular bonding) improves charge hopping and mobility.
- *Planarity* and *molecular alignment* enhance the overlap of orbitals, leading to better *charge carrier mobility*.

3.1.3.4 Effect of bonding on mechanical and thermal stability

Bonding within and between organic semiconductor molecules influences stability:

3.1.3.4.1 Effects

- *Stronger π-conjugation* may increase rigidity but reduce solubility.
- *Weak van der Waals interactions* between molecules lead to lower melting points and mechanical softness compared to inorganic semiconductors.

3.1.3.5 Bonding and chemical doping

The molecular orbitals formed by bonding determine how the material interacts with dopants:

3.1.3.5.1 Effects

- *Electron donors* raise HOMO levels; *acceptors* lower LUMO levels.
- Doping modifies charge carrier concentration and conductivity (**Table 2**).

Property	Effect of bonding
Electrical conductivity	Controlled by π-conjugation and orbital overlap
Band gap	Determined by HOMO-LUMO gap from bonding interactions
Light absorption	π-π* transitions depend on bonding extent
Charge mobility	Enhanced by delocalized bonding and planarity
Stability	Influenced by intra–/intermolecular bonding forces

Table 2.
Summary table: Effect of bonding on properties of organic semiconductors [2, 32].

3.2 Charge transport models

Charge transport in organic semiconductors differs significantly from crystalline inorganic materials due to weak intermolecular van der Waals interactions, resulting in localized states and significant disorder. Several models have been developed to describe charge carrier dynamics in these materials:

3.2.1 Hopping transport

In amorphous or polycrystalline materials, charge carriers move by thermally activated hopping between localized states. This mechanism dominates at room temperature in disordered organic semiconductors and is described by an Arrhenius-type dependence of mobility on temperature.

3.2.2 Band transport

Band transport occurs in highly ordered single-crystal organic semiconductors where delocalization is sufficient to allow carriers to propagate as Bloch waves. In this regime, mobility typically decreases with temperature, as seen in materials like rubrene [8].

3.2.3 Polaron transport

In organic materials, carriers can become coupled with lattice distortions, forming polarons. These quasi-particles move more slowly than free carriers due to the energy required to deform the lattice. Polaron formation plays a crucial role in understanding both hopping and band-like transport regimes [9].

3.2.4 Miller-Abrahams hopping model

This model quantifies hopping rates between localized states, considering both spatial separation and energy difference. The hopping probability ν_{ij} from site i to j is given by:

$$\nu_{ij} = \nu_0 \exp\left(-2\gamma r_{ij}\right) \times \begin{cases} \exp\left(-\dfrac{E_j - E_i}{kT}\right), & E_j > E_i \\ 1, & E_j \le E_i \end{cases} \qquad (2)$$

where γ is the inverse localization length, r_{ij} is the site separation, and E_i and E_j are site energies [10].

3.2.5 Multiple trap and release (MTR) model

In this model, carriers are temporarily trapped in localized states before being thermally released into the transport level. The effective mobility depends on the ratio of free to trapped carriers and is reduced by the presence of deep traps.

3.2.6 Gaussian disorder model (GDM)

Proposed by Bässler, the GDM assumes that the density of states (DOS) follows a Gaussian distribution due to energetic disorder. The model describes how disorder broadens the DOS and hinders carrier mobility, which exhibits a strong temperature dependence [11].

4. Key organic electronic devices

4.1 Organic light-emitting diodes (OLEDs)

OLEDs are electroluminescent devices that emit light in response to an electric current. They consist of a multilayer structure including a substrate, anode, hole transport layer, emissive layer, electron transport layer, and cathode. OLEDs offer wide viewing angles, high contrast ratios, and efficient color tuning. Commercial OLEDs are widely used in displays (e.g., Samsung and LG smartphones) and lighting. Research continues into enhancing efficiency, lifetime, and low-cost fabrication.

4.2 Organic photovoltaics (OPVs)

OPVs convert sunlight into electrical energy using organic semiconductors as photoactive layers. The key processes include light absorption, exciton diffusion, exciton dissociation at donor-acceptor interfaces, and charge collection. Although their power conversion efficiency (PCE) historically lagged behind silicon photovoltaics, recent innovations—such as non-fullerene acceptors—have pushed OPV efficiencies above 19% [4].

4.3 Organic field-effect transistors (OFETs)

OFETs are used in flexible logic circuits, sensors, and displays. They operate similarly to conventional MOSFETs but use an organic semiconducting layer. Key parameters include threshold voltage, mobility, and on/off current ratio. Low-voltage operation, high mechanical flexibility, and solution processability make OFETs ideal for emerging applications in wearable and disposable electronics [5].

4.4 Organic sensors

Organic materials offer high chemical sensitivity and biocompatibility, making them ideal for applications in chemical and biological sensing. Functionalized organic semiconductors can selectively detect analytes such as gases (e.g., NO_2, NH_3) or biomolecules (e.g., glucose, DNA) [6].

5. Materials and processing techniques

Organic semiconductors can be processed via several low-cost and scalable techniques:

- *Solution processing*: Including spin coating, inkjet printing, and spray coating. Suitable for large-area fabrication.
- *Vacuum deposition*: Used for small-molecule OLEDs and OFETs to create thin films with high purity.
- *Roll-to-roll printing*: Enabling continuous manufacturing of flexible electronics, akin to newspaper printing [7].

Common materials include small molecules like Alq_3 and rubrene, and polymers such as P3HT, PEDOT:PSS, and PTAA. Interface engineering and morphology control are crucial for optimizing device performance.

6. Challenges in organic electronics

Despite significant progress, several challenges remain:

- *Stability and degradation*: Organic materials are prone to degradation by oxygen, moisture, and UV light, necessitating effective encapsulation strategies.
- *Lower performance metrics*: Compared to inorganic counterparts, many organic devices still have lower carrier mobilities, efficiencies, and lifetimes.
- *Process reproducibility*: Batch-to-batch variability and sensitivity to processing conditions impact device consistency.

Addressing these challenges requires a concerted effort in materials design, interface engineering, and device architecture optimization.

7. Applications and future trends

Organic electronics offer immense potential across diverse domains:

- *Flexible and wearable electronics*: Integration into clothing and medical patches for real-time monitoring.
- *Smart packaging and labels*: Including RFID tags, electronic barcodes, and environmental sensors.
- *Bioelectronics*: Organic materials' soft and biocompatible nature makes them ideal for neural interfaces and bio-signal transducers.
- *Transparent electronics*: Organic materials can be fabricated into transparent displays and solar windows.

The future of organic electronics lies in hybrid integration with inorganic systems, sustainability through green solvents and recyclable materials, and AI-driven materials discovery.

DOI: http://dx.doi.org/10.5772/intechopen.1011119

8. Conclusions

Organic electronics have transitioned from a niche academic field to a rapidly evolving area with real-world impact. Their intrinsic advantages of flexibility, tunability, and processability open the door to novel applications that are difficult to achieve with traditional materials. While challenges such as stability and performance persist, ongoing interdisciplinary research continues to push the boundaries, promising a future where electronics are seamlessly integrated into every aspect of life. This chapter has explored the diverse applications of organic electronic devices and their potential for future advancements. With continued research and innovation, organic electronics are poised to revolutionize various industries, from energy and healthcare to flexible displays and beyond.

Acknowledgements

The authors are thankful to SERB, DST, Government of India, for the financial support under the Early Career Research Award (ECRA) for Project No. ECR/2017/000179. Authors are also thankful to VIT Vellore Management for all kinds of support. Prof. P. Chakrabarti, one of the authors, gratefully acknowledges the financial support provided by the Defense Research and Development Organization (DRDO), Government of India, through a research project (Grant No. ERIP/ER/0803699/M/01/1423).

Author details

Arun Dev Dhar Dwivedi[1*], Debashis Maji[1] and Parthasarathi Chakrabarti[2]

1 School of Electronics Engineering, Vellore Institute of Technology, Vellore, Tamil Nadu, India

2 Department of Electronics Engineering, IIT (BHU) Varanasi, Varanasi, UP, India

*Address all correspondence to: adddwivedi@gmail.com

References

[1] Shirakawa H, MacDiarmid AG, Heeger AJ. The Nobel Prize in Chemistry 2000 – Conductive Polymers. Stockholm, Sweden: The Nobel Foundation; 2000. Available from: https://www.nobelprize.org/prizes/chemistry/2000/summary/

[2] Forrest SR. The path to ubiquitous and low-cost organic electronic appliances on plastic. Nature. 2004;**428**:911-918

[3] Tang CW. Two-layer organic photovoltaic cell. Applied Physics Letters. 1986;**48**(2):183-185

[4] Yuan J et al. Single-junction organic solar cell with over 19% efficiency enabled by a refined donor–acceptor interface. Nature Energy. 2022;**7**: 561-569

[5] Horowitz G. Organic field-effect transistors. Advanced Materials. 1998;**10**(5):365-377

[6] Someya M et al. A large-area, flexible pressure sensor matrix with organic field-effect transistors for artificial skin applications. Proceedings. National Academy of Sciences. United States of America. 2004;**101**(27):9966-9970

[7] Sekitani T et al. A rubberlike stretchable active matrix using elastic conductors. Science. 2008;**321**(5895):1468-1472

[8] Podzorov V, Sysoev SE, Loginova E, Pudalov VM, Gershenson ME. Single-crystal organic field effect transistors with the hole mobility ~8 $cm^2/V{\cdot}s$. Applied Physics Letters. 2003;**83**(17):3504-3506

[9] Heeger AJ. Semiconducting and metallic polymers: The fourth generation of polymeric materials (Nobel lecture). Angewandte Chemie, International Edition. 2001;**40**(14):2591-2611

[10] Miller A, Abrahams E. Impurity conduction at low concentrations. Physics Review. 1960;**120**(3):745-755

[11] Bässler H. Charge transport in disordered organic photoconductors — A Monte Carlo simulation study. Physica Status Solidi B: Basic Solid State Physics. 1993;**175**(1):15-56

[12] Sumit Vyas ADD, Dwivedi R, Dwivedi D. Effect of gate dielectric on the performance of ZnO based thin film transistor. Superlattices and Microstructures. 2018;**120**:223-234

[13] Shubham Dadhich AD, Dwivedi D, Singh AK. Fabrication, characterization, numerical simulation and compact Modeling of P3HT based organic thin film transistors (OTFTs). Journal of Semiconductors. 2021;**42**(7):074102

[14] Dwivedi ADD, Jain SK, Dwivedi RD, Dadhich S. Numerical simulation and compact Modeling of low voltage Pentacene based OTFTs. Journal of Science: Advanced Materials and Devices. 2019;**4**(4):561-567

[15] Dwivedi ADD, Jain SK, Dwivedi RD, Dadhich S. Numerical simulation and compact modeling of thin film transistors for flexible electronics. In: INTECK Book on Hybrid Nanomaterials: Flexible Electronics Materials. UK: Inteck; 2020. DOI: 10.5772/intechopen.90301

[16] Singh MK, Anusha K, Dwivedi ADD. Enhancing device characteristics of Pentacene-based organic transistors through graphene integration: A

simulation study and performance analysis. AIP Advances. 2024;**14**:085113

[17] Anusha K, Dwivedi ADD. Comparative study of DNTT-based low-voltage BGBC, BGTC, TGBC and TGTC configurations of OTFTs. In: Book: Organic Electronics - from Fundamentals to Applications. UK: IntechOpen; 2024. DOI: 10.5772/intechopen.1006308

[18] Kaushik BK, Kumar B, Prajapati S, Mittal P. Organic Thin-Film Transistor Applications: Materials to Circuits. Boca Raton, Florida, USA: CRC Press, Taylor & Francis Group; 2017

[19] Shashi Tiwari AK, Singh L, Joshi P, Chakrabarti W, Takashima KK, Prakash R. Poly-3 hexylthiophene based organic field-effect-transistor: Detection of low concentration ammonia. Sensors and Actuators B. 2012;**171-172**:962-968

[20] Singh AK, Pandey A, Chakrabarti P. A green light sensitive Au/PBTTTC14/OTS/SiO2/Si/Ag MOS capacitor. IEEE Photonics Technology Letters. 2020;**32**:1045-1048

[21] Singh AK, Pandey A, Chakrabarti P. Poly[2,5-bis(3-tetradecylthiophen-2-yl)thieno [3,2-b]thiophene] organic polymer based-Interdigitated Channel enabled thin film transistor for detection of selective low ppm ammonia sensing at 25°C. IEEE Sensors Journal. 2020;**20**:4047-4054

[22] Srivastava A, Chakrabarti P. Fabrication and electrical characterization of a polycarbazole/ZnO based organic-inorganic hybrid heterojunction diode. Superlattices and Microstructures. 2015;**88**:723-730

[23] Srivastava A, Chakrabarti P. An organic Schottky diode (OSD) based on a-silicon/Polycarbazole contact. Synthetic Metals. 2015;**207**:96-101

[24] Dwivedi ADD, Singh AK, Prakash R, Chakrabarti P. A proposed organic Schottky barrier photodetector for application in the visible region. Current Applied Physics. 2010;**10**:900-903

[25] Singh AK, Dwivedi ADD, Chakrabarti P, Prakash R. Electronic and optical properties of electrochemically polymerized Polycarbazole/Aluminum Schottky diodes. Journal of Applied Physics. 2009;**105**:114506-1-1145065-5

[26] Srivastava A, Chakrabarti P. Experimental characterization of electrochemically polymerized Polycarbazole film and study of its behavior with different metals contacts. Applied Physics A. 2017;**123**:784

[27] Srivastava A, Chakrabarti P. Modulation of electronic conductivity and bandgap of electrochemically polymerized polycarbazole films using montmorillonite, multi-walled carbon nanotube and reduced graphene. Micro and Nano Letters. 2018;**13**:1335-1338

[28] Srivastava A, Chakrabarti P. A Novel a-Si/Polycarbazole Organic Schottky Diode (OSD) and the Method of Fabrication Thereof. Patent No. 429191; Date of Grant: 18.04.2023

[29] Heeger AJ. Semiconducting and metallic polymers: The fourth generation of polymeric materials. Journal of Physics and Chemistry of Solids. 2001;**62**(9-10):1719-1723

[30] Mac Diarmid AG. Synthetic metals: A novel role for organic polymers. Current Applied Physics. 2001;**1**(4-5):269-279

[31] Shirakawa H et al. Synthesis of electrically conducting organic

polymers: Halogen derivatives of polyacetylene. Journal of the Chemical Society, Chemical Communications. 1977;**16**:578-580

[32] Klauk H. Organic Electronics: Materials, Manufacturing, and Applications. Wiley-VCH; 2006. ISBN: 978-3527312645

[33] Facchetti A. Semiconductors for organic transistors. Materials Today. 2007;**10**(3):28-37

Chapter 2

Perspective Chapter: Charge Transport in Organic Semiconductors – Fundamentals and Improvement Methods

Jurij Urbančič

Abstract

Understanding the movement of charge carriers in organic semiconductors is essential for optimizing the performance of organic electronic devices. This chapter provides a comprehensive overview of charge transport in organic semiconductors, explaining the underlying mechanisms and discussing various transport models employed to describe mobility—the speed at which charge carriers move in response to an electric field. Furthermore, this chapter delves into the diverse methods employed to improve charge carrier mobility in organic semiconductors, such as blending the organic semiconductor with high-mobility two-dimensional material and interface engineering. Hence, this chapter presents an overview of the theoretical aspects of charge transport and various strategies for improving mobility, offering tools and insights that can contribute to the future development of organic electronic devices with enhanced performance and functionality.

Keywords: charge carrier mobility, charge transport mechanisms, organic field-effect transistor (OFET), thin films, hopping

1. Introduction

Organic semiconductor (OS) materials, including polymers and small molecules, have a significant advantage over their inorganic counterparts. They can be processed into solutions and deposited onto large, even flexible, substrates at low cost. They are also very versatile, as their physical and chemical properties can be changed by modifying their chemical structure. Because of that, they can be used to create a wide range of semiconducting electronic devices, such as light-emitting diodes, sensors, solar cells, and thin-film transistors [1, 2].

One of the most critical parameters and also a drawback of OSs is their charge mobility. Charge carrier mobility defines the velocity at which electrons and holes move through a semiconductor under an electric field. It is the ratio of drift velocity to the applied electric field. In OSs, it is several orders of magnitude lower compared to their inorganic counterparts. For example, the electron mobility of polycrystalline silicon thin

 IntechOpen

films, which is the backbone of modern electronic industry, is around 70 cm^2/Vs [3]. On the other hand, the charge carrier mobility above 10 cm^2/Vs for organic semiconductors already represents a significant advancement in research [4]. In electronic components, high mobility ensures fast signal processing and detection, as well as reduces the noise in the electrical signal and minimizes the resistive losses, thus reducing the heat generated. Thus, by optimizing the mobility of OSs, researchers can develop advanced, low-cost, and flexible devices produced using much more environmentally friendly methods compared to inorganic semiconductors.

In this book chapter, we will present a broad overview of the charge carrier dynamics in organic semiconductors. Subsequent to this introductory section, the following chapter is organized as follows: Section 2 provides the description of transport mechanism in OSs, Section 3 presents the main characterization techniques employed to determine the mobility, and Section 4 discusses various methods employed to improve the mobility of OSs. The final section provides a conclusion and perspectives. Generally, they have energy gaps ranging from 1 eV to 5 eV, are normally undoped, and have no free charge carriers while at room temperature, resulting in a low dark current [5, 6]. Only when charge carriers are injected from metallic electrodes or generated via optical excitation does the OS become conducting.

2. Transport mechanism in organic semiconductors

In solids, atoms are represented by positively charged nuclei, and the negatively charged electrons are surrounding it. According to quantum mechanics, these electrons occupy atomic orbitals, each of which has a discrete energy state. When forming a solid, enough atoms are brought close together so that the discrete energy states of the atomic orbitals begin to overlap. Due to the Pauli exclusion principle, the interaction between the electrons causes the discrete energy levels of each of the isolated atoms to split into a huge number of energy levels that are so close together that energy bands begin to form [7]. At the same time, we have the appearance of band gaps, i.e., gaps in energy between the bands. The energy band that is fully filled up with electrons is called valence band, while the energy band that has no electrons is called conduction band. The distribution of electrons over the energy states in thermal equilibrium is given by the Fermi-Dirac distribution function

$$f(E) = \frac{1}{1 + e^{\frac{E-E_F}{kT}}}, \tag{1}$$

where E_F is the Fermi level, i.e., the borderline that separates occupied and unoccupied energy levels in a material, k is the Boltzmann constant, and T is the temperature in Kelvin. The Fermi level gives us the maximum energy state that electrons occupy at absolute zero ($T = 0$). With temperatures above 0 K, the electrons just below E_F are thermally excited to levels above E_F.

Based on this theory, we can divide solids into metals, semiconductors, and insulators. At the temperature of zero Kelvin, there is an overlap of conduction and valence bands in metals, while in the case of semiconductors and insulators, they are separated, as shown in **Figure 1a**. Therefore, the E_F is located in the band gap between the valence and conduction gap. The difference between semiconductors and insulators is the width of the band gap. In semiconductors, several electrons can be

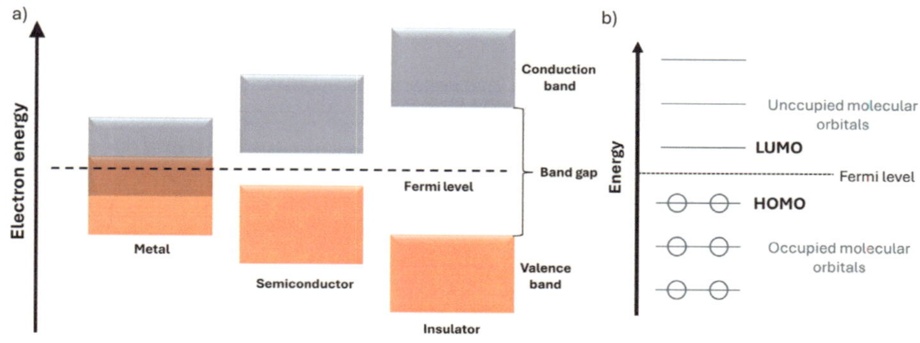

Figure 1.
(a) The band structure model explains the electrical properties of metals, semiconductors, and insulators. In metals, overlapping conduction and valence energy bands at the Fermi level allow high conductivity. Semiconductors have a small band gap, with controllable conductivity. Insulators have a large band gap, resulting in low conductivity. (b) Representation of the highest occupied molecular orbital (HOMO) and lowest unoccupied molecular orbital (LUMO).

thermally excited from valance to the conduction band, thus resulting in a measurable conductivity in the range of room temperatures. While in the case of insulators, the band gap is too large, so the probability of electrons jumping from valence band to conduction band is negligible.

The conduction mechanism describes how charge carriers (electrons and holes) can move inside the solid under the influence of an external electric field, resulting in an electrical current. Under the band theory, charge carriers can only exist and move in either the conduction or the valence band. This band transport mechanism can be used to describe inorganic crystalline semiconductors, as they have well-ordered molecular structures that form energy bands [8]. A band-like charge transport is also possible in high-purity organic molecules with a high degree of order [9], such as vacuum-deposited small molecules [10], and in single-crystal organic semiconductors such as rubrene [11, 12].

However, in the majority of organic semiconductors, due to their irregular molecular structure with many defects and trap states, the band theory does not apply. The structure of organic semiconductors is also responsible for lower charge mobility and limited current flow. Charge transport inside organic semiconductors is therefore described by energy orbital states, *LUMO* (lowest unoccupied molecular orbital) and *HOMO* (highest occupied molecular orbital), rather than energy bands, which can be seen in **Figure 1b**. The density of states (DOS) is the distribution of those energy states, representing how many energy levels are available for electrons at each energy point. In organic semiconductors, it is described by a Gaussian:

$$g(E) = \left(2\pi\sigma^2\right)^{-1/2} \exp\left(-\frac{E^2}{2\sigma^2}\right), \tag{2}$$

which is localized, and its energy (E) is centered at HOMO and LUMO. Its width σ is the energetic disorder parameter, which depends on molecular structure and impurities of the material. This means that the charge transport properties heavily depend on the morphology and local structure order within the OS. In disordered organic semiconductors, this DOS distribution applies to both charged and neutral excitations, known as excitons [13, 14]. In disordered organic semiconductors, this DOS distribution applies to both charged and neutral excitations, known as excitons.

2.1 Transport models

Several transport models have been proposed to explain charge movement in OSs, all of which must account for the structural and energetic disorder inherent in these materials. The two primary models are the hopping model and the multiple trapping and release model. **Figure 2** provides schematic representations of both models, illustrating the key mechanisms involved in charge transport.

2.1.1 Hopping

The molecular structures in most organic semiconductors, like polymers and oligomers, are highly disordered, which causes a lot of defects and traps. The energy states become localized; therefore, the energy band diagram is unable to describe the densities of states properly, and the band-like charge transport cannot explain the observed low charge mobilities, of which the value is increasing with increasing temperature, contrary to what is seen in band-like charge transport. Thus, in most organic semiconductors, the transport of charge carriers occurs via thermally activated tunneling between energetically localized states—commonly known as hopping [8].

The hopping transport in organic semiconductors is primarily described by the Miller-Abrahams model [15–17], which captures thermally activated tunneling between localized states. In simpler terms, this model accounts for charge carriers moving between states depending on their energy differences and temperature. The hopping rate can be described as a function of the separation between sites and energy levels. Similarly, the Marcus model [18, 19] applies under higher temperatures and strong electron-phonon couplings. These two models allow us to understand the dependence of mobility on temperature and electric fields without the need for band-like transport. A schematic representation of this hopping motion is shown in **Figure 2a**.

For a hopping transport, the mobility should be thermally activated, i.e., increasing with temperature. Unlike in band-like charge transport, where the mobility is described by the power law dependence on temperature, while the mobility (μ) in hopping charge transport depends on temperature and the applied electric field (F) [8]:

$$\mu(F, T) \propto \exp\left(-\frac{\Delta E}{kT}\right) \exp\left(\frac{\beta\sqrt{F}}{kT}\right), \tag{3}$$

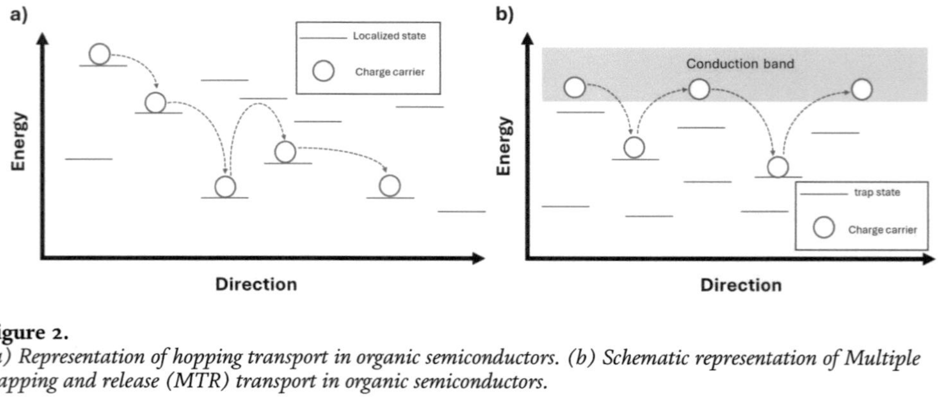

Figure 2.
(a) Representation of hopping transport in organic semiconductors. (b) Schematic representation of Multiple trapping and release (MTR) transport in organic semiconductors.

where ΔE is the activation energy and β is the disorder parameter. Charge carriers can hop either small distances with high activation energy or long distances with low activation energy.

2.1.2 Multiple trapping and release

Multiple trapping and release (MTR) [20, 21] is another model for charge transport used in amorphous organic materials to account for low mobility. Lattice defects and impurities create traps that can be classified either as shallow traps or deep traps, in which the charge carriers are immobilized. Shallow traps are located closer to valence or conduction band, while deep traps are located near the center of the band gap. The transport of carriers depends on the temperature, voltage applied, and level of trap states. Impurities and defects therefore affect the charge transport. MTR is frequently applied in well-ordered organic semiconductors while hopping is more suitable in highly disordered organic semiconductors.

The MTR working principle is based on two parts. Firstly, charge carriers are allowed to move freely in a transport band that contains delocalized energy states. And secondly, higher densities of localized energy states by the edge of the transport band are acting as traps. Charge carriers therefore move freely in the transport band while having a high probability of becoming trapped in trap states located outside of the band. If the trap in which the charge carrier gets trapped is shallow, it can be thermally released back into the transport band. The process of MTR transport is shown in **Figure 2b**. Mobility in MTR model due to the absence of localized energy states is smaller and is calculated as the effective mobility $\left(\mu_{eff}\right)$ [22]:

$$\mu_{eff} = \mu_0 \alpha \exp\left(-\frac{E_t}{kT}\right), \tag{4}$$

where E_t is the trapping energy, α is the ratio of the effective density of energy states at the edge of the transport band to the density of traps in the localized energy states, and μ_0 is the mobility in the delocalized transport band.

3. Charge mobility characterization techniques

There are several characterization methods available to determine the charge carrier mobility of an organic semiconductor. The suitability of each method depends on various factors, including the sample's thickness, the electrode's material, and the energy levels of the materials involved. Therefore, selecting the appropriate technique requires careful consideration of these specific details.

The most widely used methods are the organic field-effect transistor (OFET) method and time-of-flight photoconductivity (TOFP), which will be presented in this section. Both these experimental schematics are shown in **Figure 3**, together with representative measurements on Poly(3-hexylthiophene-2,5-diyl) (P3HT) organic semiconductor. Additional methods include the carrier extraction by linearly increasing voltage (CELIV), space-charge limited current (SCLC), photogenerated charges in CELIV (photo-CELIV), double injection (DoI), and the Hall effect measurement [9, 23].

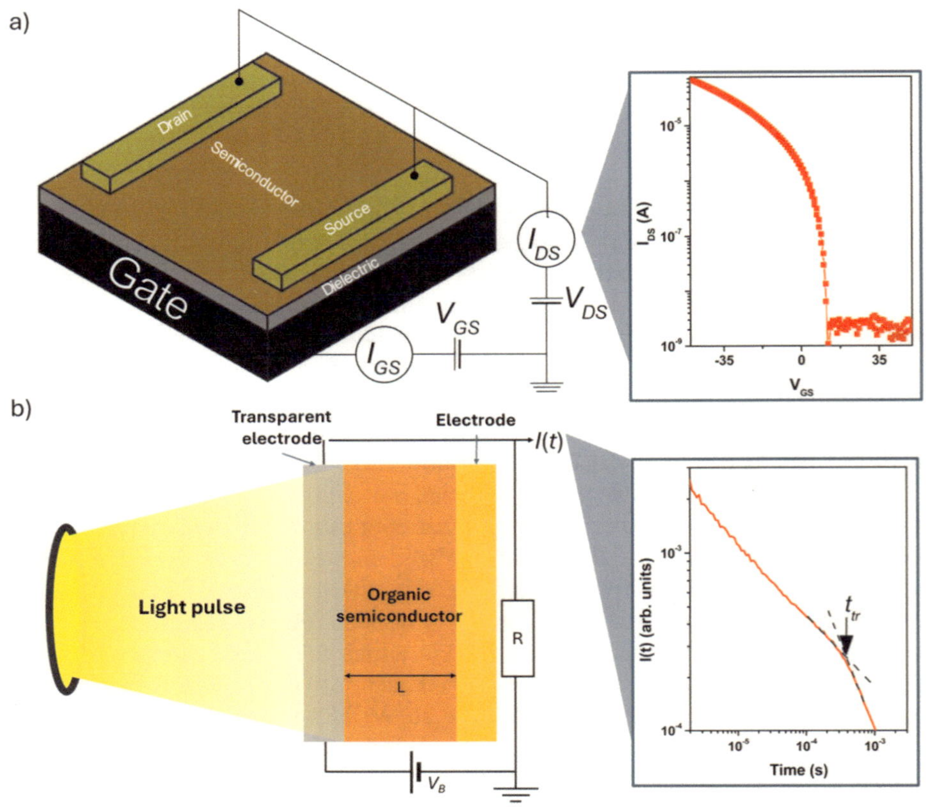

Figure 3.
(a) Schematic view of an organic field-effect transistor (OFET) for electrical characterization with a representative curve of a P3HT-based OFET. (b) Schematic representation of time-of-flight photoconductivity (TOPF) setup and representative P3HT thin-film TOFP curve. The arrow indicates the transit time t_{tr}. Both data were taken from the authors' personal measurements.

3.1 Organic field-effect transistor

This is the prevalent method for evaluating the mobility of organic semiconductors. The fabrication of organic field-effect transistors (OFETs) provides direct insight into the carrier mobility and is essential for device applications. OFETs are normal field-effect transistors (FETs), where the semiconductor layer is an organic semiconducting material.

Generally speaking, a transistor operation is analogous to the gate valve in the water pipe system. The flow from the source to drain terminals is regulated by the third terminal—the gate. In the OFET architecture, shown in **Figure 3a**, the drain, source, and gate terminals are electrodes on which we apply an external voltage. While the drain-source voltage (V_{DS}) is kept constant, the gate voltage (V_{GS}) is swept and adjusts the electric drain-source current (I_{DS}) through the semiconductor, just as the gate valve regulates the water current in the pipe system. Another important parameter is the threshold voltage V_{TH}, which is the minimum gate voltage necessary for I_{DS} to flow.

By measuring the I_{DS} as a function of V_{GS}, one can extract the charge carrier mobility. The OFET has two main modes of operation. One is where the V_{DS} is smaller than the $|V_{GS} - V_{TH}|$. When increasing the V_{DS} above the $|V_{GS} - V_{TH}|$, the transistor enters into the saturation operational regime. This case saturation measurements are generally reported when talking about charge transport in OFETs, and in this case, the mobility (μ) can be calculated according to

$$\mu = \frac{2L}{WC}\left(\frac{\partial\sqrt{I_{DS}}}{\partial V_G}\right)^2. \tag{5}$$

where W and L are the transistor channel width and length, respectively, and C is the gate capacitance per unit area. The extracted field-effect mobility is measured only at the polymer/gate dielectric interface and is prone to the effects of electrode contact resistance, traps, and defects [24]. Besides the schematic view is a typical $I_{DS}(VG)$ transconductance measurement on a P3HT-based OFET.

Another important OFET parameter is the current on/off ratio, which shows the transistor switching performance. As the name implies, it is defined as the ratio between the I_{DS} current within the transistor when it is "on" (conducting) and "off" (not conducting). For a stable and efficient OFET operation, a high on/off current ratio is desired.

3.2 Time-of-flight photoconductivity

The time-of-flight photoconductivity (TOFP) [25] is a method to characterize charge transport in organic semiconducting (OS) layers. In this case, the layer is sandwiched between two electrodes that provide an external electric field by applying a bias voltage V_B, as shown in **Figure 3c**. One electrode needs to be transparent in order for a short light (laser) pulse to be directed onto the sample in a direction perpendicular to the sample surface. The interaction between the photons (if they are of suitable energy, generally in the visible range) and the organic molecule results in the creation of excitons (electron-hole pairs) that decay in the order of nanoseconds, producing free charge carriers. Depending on the V_B polarity, either electrons or holes are transported across the sample.

The movement of photoexcited charge carriers across the OS layer due to the applied V_B gives rise to a time-dependent drift, $I(t)$, as shown in **Figure 3b** for P3HT, photoexcited with a 4 ns laser pulse with a wavelength of 532 nm. Some of the photogenerated charge carriers collide with trap states and do not participate in the charge transport, while others are released after some time. This results in a monotonically decreasing current. As the carriers reach the opposite electrode, the $I(t)$ exhibit a change in slope. This corresponds to transit time t_{tr} [26], marked in the figure by the arrow, from which the mobility can be calculated:

$$\mu = \frac{L^2}{t_{tr}V_B}, \tag{6}$$

where L is the separation between the electrodes, i.e., the length that the charge carriers travel through the OS. In this method, the charge carrier velocity is dependent only on the photoexcited charge mobility and the electric field provided by the external bias voltage. Thus, the transport properties involve paths in the entire polymer film and not only the layer at the more disordered polymer/gate dielectric interface.

4. Boosting mobility in organic semiconductors

As previously stated, the mobility of charge carriers in organic semiconductors (OSs) still presents a limitation in their applicability. Hence, great research interest is dedicated to developing new materials and fabrication methods that improve their mobility while still preserving their unique properties (such as flexibility and solubility). In this section, we will present two advanced techniques that have been developed to achieve higher mobility, e.g., the boosting of mobility by utilizing graphene nanoparticles and the utilization of a second solvent for annealing prefabricated organic thin film. Other methods also include mechanical rubbing [27, 28], blinding with small molecules [29], or carbon nanotubes [30].

4.1 Graphene-boosted mobility

Graphene is a single layer of carbon atoms arranged in a two-dimensional honeycomb lattice. Its introduction in 2004 resulted in the emerging field of 2D materials [31]. One of the most prominent characteristics of graphene is its charge mobility, with values being reported above $10^5\,cm^2/Vs$ [32]. However, the absence of an energy band gap severely limits its application to electronic devices. Therefore, its promising application is in helping to boost the mobility of OSs by combining the high charge carrier mobility of graphene and the functionality of organic materials. The explanation of this enhancement in mobility lies in the appearance of "fast lanes" for charge carriers provided by the graphene flakes within the semiconductor layer. Hence, graphene flakes are associated with the reduction of the charge carrier time spent in the semiconducting polymer.

Two key methods have been developed to enhance OS mobility, particularly in OFETs. The first involves adding graphene to the OS solution to create blends, as illustrated in **Figure 4a**. The second method is the deposition of graphene flakes at the gate dielectric/organic semiconductor (OS) interface, shown in **Figure 4b**. In the following part, we present notable examples of both techniques and their impact on device performance. A review of results in terms of mobility increase on illustrative OSs is shown in **Table 1**. As we can see, both methods achieved considerable improvement in mobility. The combination of high charge carrier mobility in graphene and the functional flexibility of organic semiconductors has thus proven to be a promising strategy for enhancing device performance.

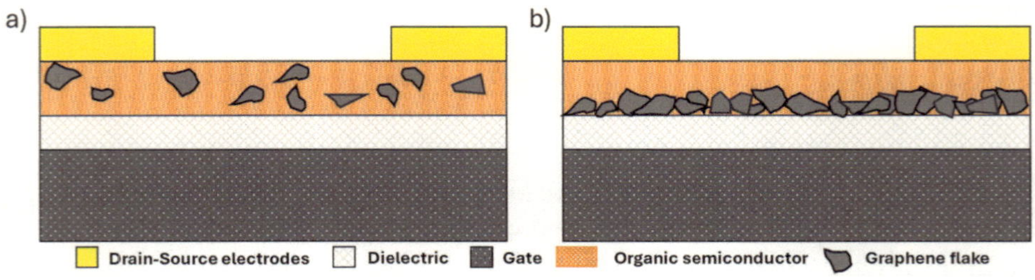

Figure 4.
Structure of the hybrid OFET made of (a) an organic semiconductor/graphene blend and (b) graphene flakes deposited on the interface between the dielectric and organic semiconductor layer.

Organic semiconductor	Method of graphene addition	Fold increase in mobility	Reference
PQT-12	OS/graphene blend	3.6[a]	[33]
P3HT	OS/graphene blend	24[a]	[33]
	OS/graphene blend	32[b]	[34]
	dielectric/OS interface	7[a]	[35]
PDIF-CN2	OS/graphene blend	2×10^{3}[a]	[36]
	OS/graphene blend	1.3×10^{3}[b]	[36]
Pentacene	dielectric/OS interface	2[a]	[35]

[a]*OFET measurements.*
[b]*TOFP measurements.*

Table 1.
A review of results when combining graphene flakes with the organic semiconductor (OS) is presented here in terms of mobility increase.

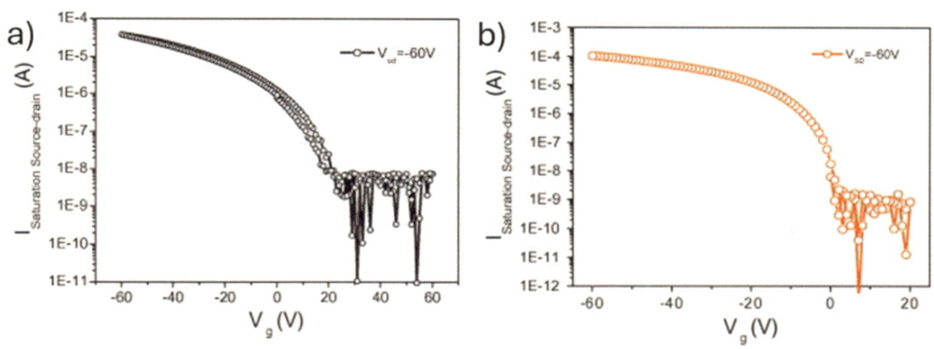

Figure 5.
OFETs transfer curves of (a) neat PQT-12 and (b) PQT-12/graphene blend. Reproduced with permission from Ref. [33]. Copyright 2011 Elsevier B.V.

4.1.1 Organic semiconductor: Graphene blends

In the case of PQT-12 [33], the transfer curves for both pristine PQT-12 and PQT-12:graphene blend are shown in **Figure 5**. From them, the effective mobility of the neat PQT-12 was estimated to be 0.15 cm^2/Vs. In contrast, the effective mobility of the PQT-12/graphene blend is 0.55 cm^2/Vs, representing a 3.6-fold increase in mobility due to the addition of graphene.

The P3HT:graphene blend exhibits field-effect mobility as high as 0.17 cm^2/Vs, an impressive improvement from that of neat P3HT ($7 \times 10^{-3} cm^2/Vs$) [33], corresponding to a 24-fold increase. A comparison of their transconductance curves is shown in **Figure 6a**. This improvement is further confirmed with TOFP measurements [34], where the charge mobility increases from 4×10^{-5} (pristine P3HT) to $1.3 \times 10^{-3} cm^2/Vs$, a 32-fold increase. The measured $I(t)$ curves are shown in **Figure 6b**.

With respect to PDIF-CN2 [36], both OFET and TOFP measurements indicate enhanced electron mobility with this method, as shown in **Figure 7**. The OFET data

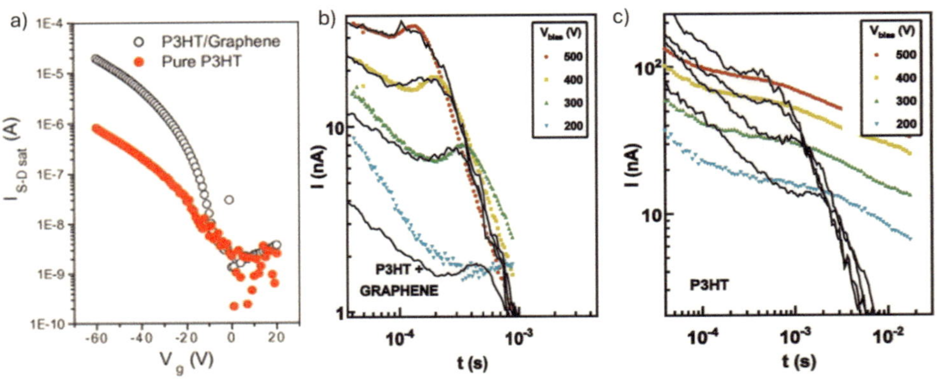

Figure 6.
OFET transfer characteristics of (a) pure P3HT and (b) P3HT:graphene blend. Reproduced with permission from Ref. [33]. Copyright 2011 Elsevier B.V. The measured I(t) curves from the time-of-flight photoconductivity (TOFP) measurement on (c) P3HT:graphene blend and (d) neat P3HT. Reproduced with permission from Ref. [34]. Copyright 2019 Elsevier Ltd.

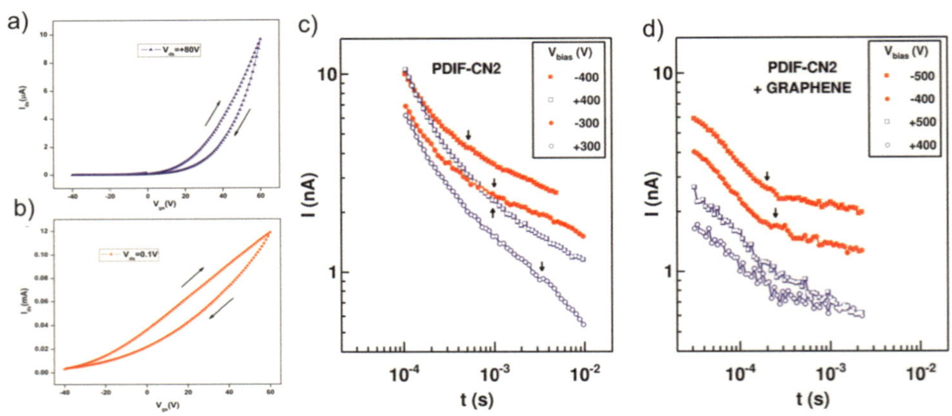

Figure 7.
Typical transfer characteristics of a thin-film transistor based on (a) a neat PDIF-CN2 layer and (b) a graphene blend. Plot of the time-of-flight photoconductivity current I(t) of (c) PDIF-CN2 and (d) PDIF-CN2:graphene blend. Reproduced with permission from Ref. [36]. Copyright © 2014 American Chemical Society.

(**Figure 7a**) show pristine PDIF-CN2 has a mobility of 10^{-4} cm^2/Vs, while the PDIF-CN2:graphene blend exhibits a remarkable 0.2 cm^2/Vs, a 2×10^3-fold increase. TOFP results (**Figure 7b**) also confirm improved charge carrier mobility, increasing from 1.3×10^{-4} cm^2/Vs to 9.0×10^{-4} cm^2/Vs with a 530 nm laser. A 210 nm laser further boosts mobility in the blend to 0.17 cm^2/Vs (1.3×10^3-fold increase), consistent with OFET measurements. This wavelength dependence is attributed to charge carrier generation in PDIF-CN2 at lower photon energies (510 nm) and hot electron excitation in graphene flakes at higher energies (210 nm).

4.1.2 Graphene flakes at the gate dielectric/organic semiconductor interface

The results for OFETs with graphene flakes at the interface are presented for the case of P3HT and pentacene [35]. Regarding P3HT (**Figure 8a**), samples without graphene flakes achieved field-effect mobility of 7×10^{-3} cm^2/Vs, while the samples

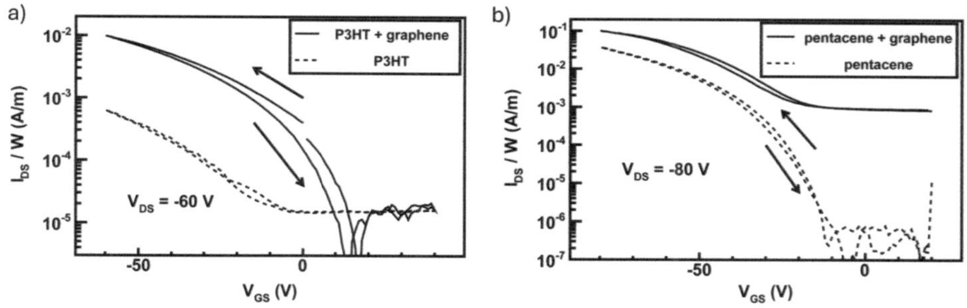

Figure 8.
Comparison of the transfer characteristics on a plot with (solid curve) and without (dashed curve) graphene flakes for (a) P3HT-based and (b) pentacene-based OFETs. Reproduced with permission from Ref. [35]. Copyright 2015 Elsevier B.V.

with the graphene flakes exhibited saturated mobility of 5×10^{-2} cm^2/Vs, a 7-fold increase. For pentacene, the mobility extracted from transfer curves (**Figure 8b**) has increased from 0.16 to 0.37 cm^2/Vs for graphene-coated dielectric, a 2-fold increase, although this is also accompanied by a drop in on/off ratio.

4.2 Solvent annealing

Solvent annealing is widely used to improve the performance of organic semiconductor films. During this process, a pre-deposited organic film is exposed to a solvent, often in vapor form, which causes the film to swell. This swelling allows the molecules to rearrange into a more ordered configuration. Once the solvent evaporates, the film solidifies with increased crystallinity and larger grain size, leading to fewer grain boundaries and better charge transport. These structural improvements significantly enhance the mobility of charge carrier transport [37–39]. Solvent annealing is particularly valued for its ability to fine-tune the morphology of organic films, making it a critical step in the fabrication of high-performance OFETs and other organic electronic devices.

To illustrate the effectiveness of this technique, we will now explore the following notable examples: P3HT [40]; triethylsilylethynyl anthradithiophene (TES ADT) [41]; Copper(II) phthalocyanine (CuPc) [42]. These examples demonstrate how solvent annealing can significantly enhance the structural properties and charge transport capabilities of different organic semiconductor materials. **Table 2** summarizes their increase in mobility using this method.

For P3HT [40], exposing spin-coated films to dichlorobenzene vapor increases the field-effect mobility from 1.3×10^{-3} to 1.2×10^{-3} cm^2/Vs under unsaturated vapor conditions[1]. Atomic force microscopy (AFM) reveals that fiber-like structures, characteristic of crystalline P3HT, are indistinct in the sample without vapor annealing (**Figure 9a**) but become visible after unsaturated vapor treatment (**Figure 9b**). Under saturated vapor annealing,[2] mobility increases significantly to 0.11 cm^2/Vs. The AFM image in **Figure 9c** shows the P3HT film absorbed by the solvent vapor, leaving an estimated 2 nm-thin layer (based on scans across a scraped line at the top right corner).

[1] Annealing occurred in a partially sealed volume.

[2] Annealing took place in a fully sealed volume filled with saturated DCB vapor.

Organic semiconductor	Solvent	Fold increase in mobility	Reference
P3HT	Chlorobenzene	9[a]	[40]
		84[b]	
TES ADT	Acetone	5	[41]
	1,2-Dichloroethylene	55	
	Hexanes	1	
	Tetrahydrofuran	10	
	Toluene	25	
Copper(II) phthalocyanine	Acetone	7	[42]
	IPA	11	

[a]*Unsaturated-vapor-annealed sample.*
[b]*Saturated-vapor-annealed sample.*

Table 2.
A review of mobility increases using solvent annealing technique on different organic semiconductors.

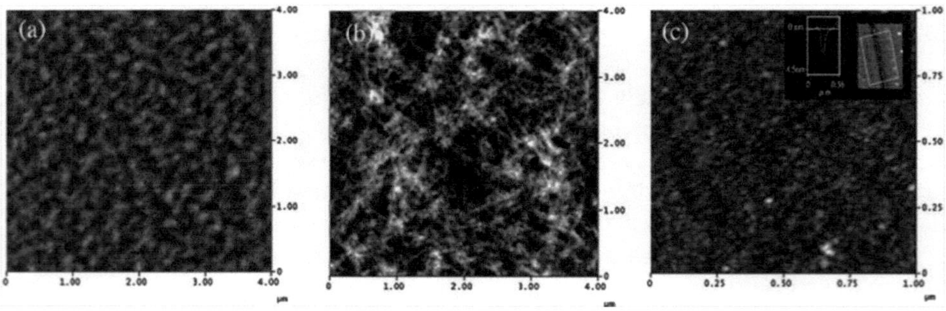

Figure 9.
(a) AFM topography images of P3HT films without solvent vapor annealing, (b) annealed in unsaturated solvent vapor, and (c) annealed with saturated solvent vapor—in the top right corner is thickness as determined from the scan made over a scraped line through the film. Reproduced with permission from Ref. [40]. Copyright © 2009 Elsevier B.V. Modifications include the merging of images.

The vapor annealing process has been shown to dramatically improve the electrical properties of triethylsilylethynyl anthradithiophene (TES ADT) [41]. Solvents of varying polarity were tried, most showing great improvement in terms of mobility increase, as shown in **Figure 10** after 2 minutes of exposure to time. Using 1,2-Dichloroethane for solvent annealing yields the highest mobility (μ) results, resulting in a jump from 2×10^{-3} to 1.1×10^{-1} cm^2/Vs—a 55-fold increase. The OFET characteristics, in addition to μ that were also investigated, are the change in the current on/off ratio and the shift in threshold voltage (V_T). As a result, it was demonstrated that OFETs fabricated with TES ADT and then subjected to solvent vapor annealing exhibited significantly enhanced device characteristics, meeting the performance standards required for practical applications.

OFETs fabricated from CuPC and treated with acetone and isopropyl alcohol (IPA) have also shown a higher transistor performance. In this case, both the morphological changes and the mobility performance were investigated. The morphology of the organic thin film was estimated using the X-ray diffraction analysis (XRD) and AFM

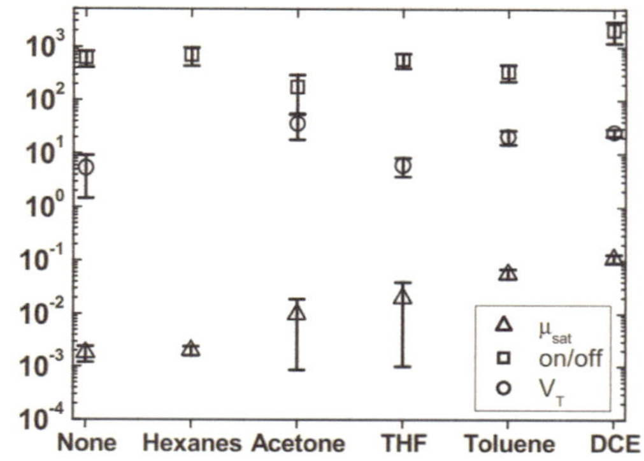

Figure 10.
TES ADT OFET device characteristics after 2 minutes of solvent vapor annealing of different solvents. In addition to the mobility (μ), the on/off current ratio and threshold voltage (V_T) are presented. Reproduced with permission from Ref. [41].

(see **Figure 11**). XRD data (**Figure 11a**) was used to estimate the crystallinity of the film. In the case of pristine CuPC, the crystallinity was observed to be 14.5%. For acetone- and IPA-treated samples, the crystallinities were found to be 18% and 20%, correspondingly. In the case of the average grain size, their value was extracted from the AFM images (**Figure 11b–d**). For the pristine case, it was observed to be in the range between 90 and 110 nm (**Figure 11b**). For acetone treatment, the average grain

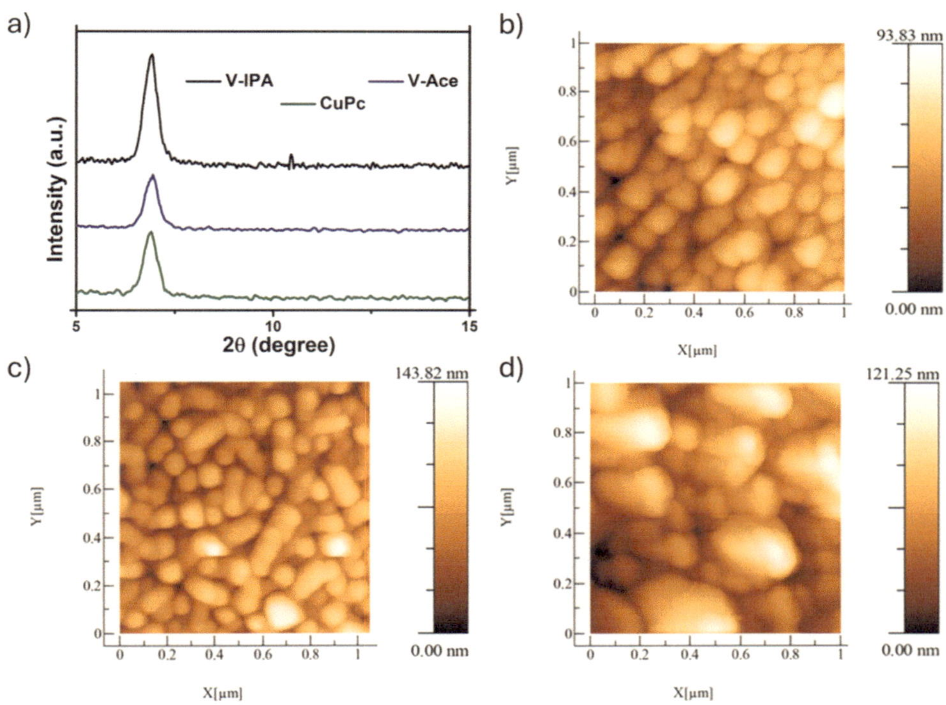

Figure 11.
Morphology measurements on CuPc films. (a) XRD data on untreated and treated samples. (b–d) AFM images of CuPC films in the cases of (b) untreated CuPc and solvent vapor annealment with (b) acetone and (c) IPA. Reproduced with permission from Ref. [42]. Copyright © 2021 Elsevier Ltd. Modification of the images includes merger and removal of excessive data from image (a).

size increases to the range between 150 and 250 nm (**Figure 11c**). The largest grains were found in the IPA-treated CuPC films, with the average size being in the range between 350 and 440 nm (**Figure 11d**).

In an untreated CuPC OFET, the mobility is estimated at $2.3 \times 10^{-4}\,cm^2/Vs$. After solvent vapor annealing with acetone, the mobility improves to $1.7 \times 10^{-3}\,cm^2/Vs$, representing a 7-fold increase. Treatment with IPA further enhances mobility to $2.6 \times 10^{-3}\,cm^2/Vs$, an 11-fold increase. Notably, the current on/off ratio also rises from 10^2 to 10^3 following solvent vapor annealing. These performance improvements align with the results of the morphology investigation, confirming the relationship between crystallinity and grain size with the enhancement in charge carrier mobility.

5. Conclusion and future perspectives

In this chapter, we have explored the fundamentals of charge transport in organic semiconductors, highlighting the mechanisms that govern it and presenting two advanced techniques employed to enhance the mobility of charge carriers. Despite significant advances being made in the last decades, the mobility of charge carriers in organic semiconductors remains a limiting factor in their application in electronic devices. Techniques such as graphene incorporation and solvent annealing have demonstrated remarkable improvements in mobility of up to a couple of orders of magnitude, thereby pushing the boundaries of organic electronics. The successful integration of graphene with organic semiconductors has provided new pathways for faster charge transport, while solvent annealing has refined the morphological and crystalline structures, leading to more efficient charge movement.

Looking forward, the continuous evolution of material science and nanotechnology is expected to further overcome the limitations of organic semiconductors. Future research should focus on the development of new materials with tailored properties, advanced fabrication techniques, and possible hybrid systems that combine the strengths of organic and inorganic components. In conclusion, the goal is to achieve higher mobility, stability, and functionality in organic electronic devices, paving the way for their widespread application in next-generation electronic devices.

Conflict of interest

The authors declare no conflict of interest.

Author details

Jurij Urbančič
University of Nova Gorica, Nova Gorica, Slovenia

*Address all correspondence to: jurij.urbancic@ung.si

References

[1] Kymissis I. Organic Field Effect Transistors: Theory, Fabrication and Characterization. New York, USA: Springer Science & Business Media; 2008

[2] Ostroverkhova O. Handbook of Organic Materials for Electronic and Photonic Devices. Sawston, United Kingdom: Woodhead Publishing; 2018

[3] Hawkins WG. Polycrystalline-silicon device technology for large-area electronics. IEEE Transactions on Electron Devices. 1986;**33**(4):477-481

[4] D'Amico F, He X, Jiang H. Greater than 10 $cm^2\ V^{-1}\ s^{-1}$: A breakthrough of organic semiconductors for field-effect transistors. InfoMat. 2021;**3**(6):613-630

[5] Smith DL. Electronic properties of inorganic and organic semiconductors and their application to national security needs. MRS Bulletin. 2004;**29**(9): 647-652

[6] Scholz R. Organic semiconductors. Encyclopedia of Condensed Matter Physics. Vol. 4. 2005. p. 206

[7] Holgate SA. Understanding Solid State Physics. Boca Raton, USA: CRC Press; 2021

[8] Li W, Kwok H. Conduction mechanisms in organic semiconductors. In: Encyclopedia of Nanotechnology. Dordrecht, Netherlands: Springer; 2012

[9] Karl N. Charge carrier transport in organic semiconductors. Synthetic Metals. 2003;**133**:649-657

[10] Ji D, Li T, Liu J, Amirjalayer S, Zhong M, Zhang ZY, et al. Band-like transport in small-molecule thin films toward high mobility and ultrahigh detectivity phototransistor arrays. Nature Communications. 2019;**10**(1):1-8

[11] Podzorov V, Menard E, Borissov A, Kiryukhin V, Rogers JA, Gershenson ME. Intrinsic charge transport on the surface of organic semiconductors. Physical Review Letters. 2004;**93**(8):086602

[12] Xie W, McGarry KA, Liu F, Wu Y, Ruden PP, Douglas CJ, et al. High-mobility transistors based on single crystals of isotopically substituted rubrene-d 28. The Journal of Physical Chemistry C. 2013;**117**(22): 11522-11529

[13] Bässler H. Charge transport in disordered organic photoconductors, a Monte Carlo simulation study. Physica Status Solidi B (Basic Research), Germany. 1993;**175**(1):18-19

[14] Fishchuk II, Kadashchuk A, Hoffmann ST, Athanasopoulos S, Genoe J, Bassler H, et al. Unified description for hopping transport in organic semiconductors including both energetic disorder and polaronic contributions. Physical Review B. 2013; **88**(12):125202

[15] Bässler H, Schönherr G, Abkowitz M, Pai DM. Hopping transport in prototypical organic glasses. Physical Review B. 1982;**26**(6):3105

[16] Borsenberger PM, Richert R, Bässler H. Dispersive and nondispersive charge transport in a molecularly doped polymer with superimposed energetic and positional disorder. Physical Review B. 1993;**47**(8):4289

[17] Pavlica E, Bratina G. The role of space-charge-induced electric field on transient photocurrent response in

organic semiconductors. Physica Status Solidi (B). 2005;**243**(2):473-482

[18] Coropceanu V, Cornil J, da Silva Filho DA, Olivier Y, Silbey R, Bredas JL. Charge transport in organic semiconductors. Chemical Reviews. 2007;**107**(4):926-952

[19] Lu N, Li L, Banerjee W, Sun P, Gao N, Liu M. Charge carrier hopping transport based on Marcus theory and variable-range hopping theory in organic semiconductors. Journal of Applied Physics. 2015;**118**(4):045701

[20] Nenashev AV, Oelerich JO, Baranovskii SD. Theoretical tools for the description of charge transport in disordered organic semiconductors. Journal of Physics: Condensed Matter. 2015;**27**(9):093201

[21] Dey A, Singh A, Das D, Iyer PK. Organic Semiconductors: A New Future of Nanodevices and Applications. Cham, Switzerland: Springer; 2015

[22] Pei M, Guo J, Zhang B, Jiang S, Hao Z, Xu X, et al. Semiconductor/dielectric interface in organic field-effect transistors: Charge transport, interfacial effects, and perspectives with 2D molecular crystals. Advances in Physics: X. 2020;**5**(1):1747945

[23] Kokil A, Yang K, Kumar J. Techniques for characterization of charge carrier mobility in organic semiconductors. Journal of Polymer Science Part B: Polymer Physics. 2012;**50**(15):1130-1144

[24] Chang J, Lin Z, Zhang C, Hao Y. Organic field-effect transistor: Device physics, materials, and process. In: Different Types of Field-Effect Transistors-Theory and Applications. Rijeka, Croatia: InTech; 2017. p. 22

[25] Bratina G, Pavlica E. Characterisation of charge carrier transport in thin organic semiconductor layers by time-of-flight photocurrent measurements. Organic Electronics. 2019;**64**:117-130

[26] Pavlica E, Bratina G. Time-of-flight mobility of charge carriers in position-dependent electric field between coplanar electrodes. Applied Physics Letters. 2012;**101**(9):093304

[27] Heil H, Finnberg T, Von Malm N, Schmechel R, von Seggern HV. The influence of mechanical rubbing on the field-effect mobility in polyhexylthiophene. Journal of Applied Physics. 2003;**93**(3):1636-1641

[28] Derue G, Serban DA, Leclère P, Melinte S, Damman P, Lazzaroni R. Controlled nanorubbing of polythiophene thin films for field-effect transistors. Organic Electronics. 2008;**9**(5):821-828

[29] Orgiu E, Masillamani AM, Vogel JO, Treossi E, Kiersnowski A, Kastler M, et al. Enhanced mobility in P3HT-based OTFTs upon blending with a phenylene–thiophene–thiophene–phenylene small molecule. Chemical Communications. 2012;**48**(10):1562-1564

[30] Bo XZ, Lee CY, Strano MS, Goldfinger M, Nuckolls C, Blanchet GB. Carbon nanotubes-semiconductor networks for organic electronics: The pickup stick transistor. Applied Physics Letters. 2005;**86**(18):182102

[31] Novoselov KS, Geim AK, Morozov SV, Jiang DE, Zhang Y, Dubonos SV, et al. Electric field effect in atomically thin carbon films. Science. 2004;**306**(5696):666-669

[32] Bolotin KI, Sikes KJ, Jiang Z, Klima M, Fudenberg G, Hone J, et al.

Ultrahigh electron mobility in suspended graphene. Solid State Communications. 2008;**146**(9–10):351-355

[33] Huang J, Hines DR, Jung BJ, Bronsgeest MS, Tunnell A, Ballarotto V, et al. Polymeric semiconductor/graphene hybrid field-effect transistors. Organic Electronics. 2011;**12**(9):1471-1476

[34] Pathipati SR, Pavlica E, Bratina G. Enhanced photoconductivity of semiconductor polymers using graphene nanoflakes. Materials Today: Proceedings. 2019;**19**:2591-2595

[35] Pathipati SR, Pavlica E, Parvez K, Feng X, Müllen K, Bratina G. Graphene flakes at the SiO_2/organic-semiconductor interface for high-mobility field-effect transistors. Organic Electronics. 2015;**27**: 221-226

[36] Pathipati SR, Pavlica E, Schlierf A, El Gemayel M, Samori P, Palermo V, et al. Graphene-induced enhancement of n-type mobility in perylenediimide thin films. The Journal of Physical Chemistry C. 2014;**118**(43):24819-24826

[37] Yadav S, Kumar P, Ghosh S. Optimization of surface morphology to reduce the effect of grain boundaries and contact resistance in small molecule based thin film transistors. Applied Physics Letters. 2012;**101**(19):193307

[38] Xu J, Liu X, Hou W, Guo H, Yu L, Zhang H. Effect of solvent–vapour annealing on morphology, structure of copper (II) phthalocyanine thin films and device performance. Bulletin of Materials Science. 2018;**41**:1-7

[39] Kösemen ZA, Kösemen A, Öztürk S, Canımkurbey B, Erkovan M, Yerli Y. Performance improvement in photosensitive organic field effect transistor by using multi-layer structure. Thin Solid Films. 2019;**672**:90-99

[40] Fu Y, Lin C, Tsai FY. High field-effect mobility from poly (3-hexylthiophene) thin-film transistors by solvent–vapor-induced reflow. Organic Electronics. 2009;**10**(5):883-888

[41] Dickey K, Anthony J, Loo YL. Improving organic thin-film transistor performance through solvent-vapor annealing of solution-processable triethylsilylethynyl anthradithiophene†. Advanced Materials. 2006;**18**:1721-1726

[42] Kumar P, Yadav S, Kumar N, Kumar L. Effect of organic solvent vapor treatment on transistor performance and contact resistance of copper phthalocyanine based organic field-effect transistors. Solid-State Electronics. 2021;**176**:107954

Chapter 3

Towards Organic Electronics: Enhanced PEDOT:PSS Electrode Conductivity

Aaron DiFilippo, Amrita Chakraborty and Marius K. Orlowski

Abstract

This chapter focuses on enhancing the electrical conductivity of Poly (3,4-ethylenedioxythiophene) Polystyrene Sulfonate (PEDOT:PSS:PSS) films for flexible substrate-based organic electronics. Techniques such as acid treatments, doping with noble and non-noble metal nanoparticles (Cu and Ag), multiple layer depositions, and the incorporation of mono/multiatomic layer graphene are systematically investigated. The optimized process prioritizes PEDOT:PSS multilayer depositions and nitric acid treatment, proving more effective and cost-efficient compared to alternatives using metal nanoparticles and graphene. When exposed to nitric acid, the optimized multilayer PEDOT:PSS:PSS stack decreases sheet resistance to 7.66 Ω/sq. from over 1 MΩ/sq., which results in an extraordinary conductivity of 15,701 S/cm, as opposed to 0.17 S/cm for single PEDOT:PSS layer films. The research details adhesion techniques using oxygen plasma treatment and patterning methods involving Ag sacrificial layers. The enhanced electrical conductivity achieved through multiple PEDOT:PSS: PSS depositions does not result in a significant increase in thickness, allowing for the attainment of very high electrical conductivity values. The study also addresses film aging and strategies to mitigate reliability effects induced by ambient conditions. These findings offer significant advancements in the fabrication of organic conductive electrodes, providing engineering flexibility for various applications and ensuring long-term reliability and performance of the films.

Keywords: PEDOT: PSS, conductive polymers, flexible organic electronics, doping w/metal nanoparticles, acid treatment, insertion of mono- and tri-layer graphene

1. Introduction

Poly (3,4-ethylene dioxythiophene): polystyrene sulphonate (PEDOT:PSS) [1] stands out as a highly promising option for organic conductive films. Alongside polyacetylene, polypyrrole (PPy), polyaniline (PANI), and P3HT, it has garnered significant interest across various application areas. The advantages of PEDOT:PSS are manifold: it is lightweight; easily processed and deposited through methods like spin coating, spray coating, and ink-jet printing; possesses good mechanical flexibility and integrity; and exhibits thermal stability. Moreover, it has the potential to increase conductivity by several orders of magnitude, with a work function suitable for use in hybrid crystalline silicon solar cells [2]. PEDOT:PSS finds applications in energy devices, next-generation photovoltaics [3], optoelectronic devices as transparent and flexible electrodes [4–9], and functional packaging layers [10–12]. It has effectively replaced indium-tin oxide (ITO) films [13–15], serves as a buffer layer in charge extraction and injection devices [16–18], and is utilized in energy storage cells [19, 20], thermoelectric materials [21], bioelectronic devices [22, 23], and ITO-free liquid crystal displays [24]. Highly conductive PEDOT:PSS electrode also has the potential to replace Au electrodes in organic thin film transistors (OTFT) [25, 26]. Despite these advantages, the electrical conductivity of pristine PEDOT:PSS films remains relatively low, around 0.1 S cm^{-1} [24]. Various techniques have been developed to enhance conductivity, including increasing PEDOT grain concentration, blocking electrostatic attraction between PEDOT and PSS, and using nanoparticles or nanocomposites to fill spaces between PEDOT islands within the bulk of PSS [1]. Patterning of conductive polymers is crucial for creating organic electrodes, achieved through techniques like ink-jet printing, vapor deposition, imprint lithography, and photolithography [27–38]. Recent advancements in photolithography methods have addressed challenges such as chemical incompatibility, resulting in successful PEDOT:PSS patterning [34–38]. Our earlier work [39, 40] introduced a method for manufacturing conductive organic electrodes using doped PEDOT-PSS polymer sheets, demonstrating strong substrate adherence and substantial conductivity improvement through successive depositions. However, combining multiple approaches, such as Cu NP doping and multilayer deposition, resulted in counteractive effects [39, 40].

In this chapter, we describe new methods building upon our previous work, including mono- and multi-layer graphene, and addition of noble metal nanoparticles, and acid treatments. Among acids, nitric acid treatment performs best overall, showing promising results for enhancing conductivity [41–45]. In earlier studies, the addition of metal nanoparticles to PEDOT:PSS has attracted a lot of interest. For example, Zhang et al. [41] effectively created PEDOT:PSS thin films doped with silver and copper nanoparticles using inkjet printing, resulting in excellent optical and electrical properties. Similarly, Patil et al. [42] found that doping PEDOT:PSS with silver nanoparticles increased its conductivity. Zhang et al. [43] demonstrated the effectiveness of gold nanoparticle-PEDOT:PSS nanocomposites as catalysts in alkaline direct ethanol fuel cells by providing a simplified synthesis procedure for their fabrication. By using silver nanoparticles (Ag NP), Ghazy et al. [44] improved the PEDOT:PSS conductivity utilized in organic solar cells. Finally, strong evidence of significant increases in polymer conductivity following the application of Cu NPs as dopants was presented by Pham et al. [45]. Adding graphene (in monolayer and triple-layer versions) offers advantages of its own, just like with metal nanoparticles; nevertheless, in contrast to silver nanoparticles, it does not work in concert with nitric acid treatment. However, conductivity deterioration is caused by ambient-induced effects and film aging, which can be addressed by adjusting the nitric acid concentration. When compared to methods that use metal nanoparticles and graphene layers, the straightforward process that combines numerous PEDOT:PSS depositions with appropriate nitric acid treatment provides an easy and affordable way to boost PEDOT:PSS conductivity. Sheet resistance is reduced from 1 MΩ/sq. to 6.66 Ω/sq. in the optimized multilayer PEDOT:PSS treated with optimum nitric acid, representing significant improvement compared to previous methods [39, 40]. In addition, phosphoric acid treatment shows promise for enhancing solar cell efficiency [46–48]. Lastly, the research group led by Xie [41] demonstrated the creation of highly conductive PEDOT:PSS transparent electrodes for polymer solar cells using a post-spin-rinsing technique. In previous studies, graphene has been incorporated into PEDOT:PSS films in various forms to enhance their electrical conductivity significantly. One study [49] integrated graphene composites into PEDOT:PSS films for energy harvesting applications by dispersing the composites in a PSS solution and adjusting their concentration to optimize conductivity. This method resulted in a 41% increase in conductivity compared to undoped PEDOT:PSS, leading to a 93% improvement in power factor relative to devices using pristine PEDOT:PSS.

In another study [50], graphene was used to dope PEDOT:PSS films, following a similar approach involving treatment with concentrated H_2SO_4 to enhance electrocatalytic activity at fuel cell and metal-air battery cathodes. The resulting PEDOT:PSS/graphene composites demonstrated enhanced electrocatalytic activity, increased resistance to the methanol crossover effect and CO poisoning, and greater durability compared to Pt/C electrodes. Additionally, Mahakul et al. [51] employed reduced graphene oxide-carbon nanotube doping to improve the conductivity of PEDOT:PSS films for transparent electrode applications, achieving a maximum conductivity of 3804 S/cm, comparable to that of indium oxide (4000 S/cm). In reference [52], PEDOT:PSS films were doped with graphene and graphene quantum dots. Results showed that both dopants enhance the conductivity of PEDOT:PSS films while maintaining high transparency, with only a slight reduction in transparency (13-14%).

Liu et al. created durable, highly conductive polymer films for biosensor applications by doping PEDOT:PSS with graphene nanocomposites in a related work [53]. They used an electrospray method to deposit graphene nanoplatelet (GNP) composites atop fluoride tin oxide (FTO), and then they treated the material with H_2SO_4. The improved FTO electrode effectively distinguished between the electrochemical oxidation signals of ascorbic and uric acids, demonstrating strong catalytic activity in the detection of dopamine. A flexible sensor composed of graphene oxide/PEDOT:PSS composites was introduced by Ko et al. [54] for the voltametric detection of low dopamine levels. The sensor achieved a sensitivity of 69.3 $\mu A/\mu Mcm^2$ and a detection limit of 0.008 μM.

Further advancements were reported by Du et al. [55], who incorporated graphene quantum dots (GQDs) into PEDOT:PSS to attain high-conductivity polymer films. They increased the thermal conductivity by 113% and the electrical conductivity by 717 S/m, which is significantly better than the pristine PEDOT:PSS. A simple casting technique was used to create the PEDOT:PSS/GQDs. Strong π-π bonding between GQDs and PSS chains caused phase separation and decoupling, which enhanced the electrical and thermal conductivities. Exfoliated graphene and PEDOT:PSS were mixed by Badri et al. [56] as an alternative to indium tin oxide for optoelectronic devices, achieving a balance between conductivity (up to 4.2×10^3 S/cm) and optical transparency (up to 94%). In another study [57], PEDOT:PSS and graphene were used to fabricate an electrochromic device with active, flexible conductive electrode films, utilizing the wrinkling instability of graphene to impart hydrophobic properties, resulting in a broad color contrast range, flexibility, and anti-wetting characteristics. A thorough analysis of different PEDOT:PSS doping methods using graphene and its derivatives, such as graphene oxide, reduced graphene oxide, and graphene quantum dots, was presented by Adekoya et al. [58]. This review highlighted recent developments in combining PEDOT:PSS with graphene derivatives for energy storage device applications, such as supercapacitors [59].

The structure of this chapter comprises several sections, beginning with electrode fabrication, followed by acid treatment procedures, graphene and metal nanoparticle insertion, and individual technique results. The paper then delves into the synergistic benefits of these processes, concluding with remarks that summarize the key findings.

2. Polymer electrode fabrication

This article highlights innovative strategies to enhance the electrical conductivity of PEDOT:PSS. The device fabrication process is thoroughly detailed in our earlier studies [39, 40]. We investigate the use of acid treatments with nitric, sulfuric, and phosphoric acids, as well as the incorporation of silver and copper nanoparticles, both within the material and as surface treatments. We also investigate the effects of varying environmental conditions on the organic electrodes by keeping them for up to 10 days in vacuum chambers and conventional laboratory settings. Prior to exploring these novel enhancing methods, it is helpful to provide an overview of the electrode fabrication process, which is addressed in detail in Refs. [39, 40]. To create organic electrodes, a commercial PEDOT:PSS dispersion in water was first used. To create organic films, this dispersion was applied to clean substrates like flexible Mylar substrates and oxidized silicon wafers. For film deposition, we employed the spin-coating process, varying the spin speed from 500 to 3000 rpm. To obtain the necessary film thickness and minimize sheet resistance, the spin speed selection was critical for the development of multilayer PEDOT:PSS stacks, where the speed of each layer, including speed ramps, was carefully controlled. Furthermore, oxidized silicon wafer was cleaned using oxygen plasma, which greatly enhanced the films' uniformity. To precisely measure the film thickness, the PEDOT:PSS layers need to be patterned to achieve a sharp step. Atomic Force Microscopy (AFM) and a Dektak profilometer were used to measure the thickness of PEDOT:PSS films, and the results were within +/−1 nm of one another. An example of thickness measurement by AFM is shown in **Figure 1**.

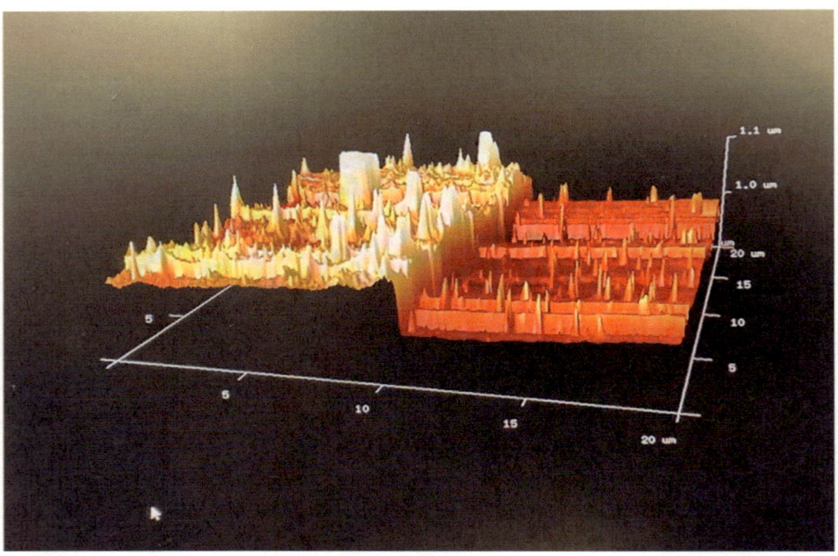

Figure 1.
An image of measurement of the PEDOT:PSS film thickness by AFM. In addition to the step of the patterned PEDOT:PSS film, the surface roughness of the PEDOT:PSS is shown.

PEDOT:PSS #×	1 × (a)	1 × (b)	1 × (c)	3 × (a)	3 × (b)	6×	9×
Thickness [nm]	65	56	29	85	72	113	146

The spin speeds for each individual layer are as follows: Single layer 1 × (a) at 1500 rpm, single layer 1 × (b) at 2000 rpm, and single layer 1 × (c) at 3000 rpm. For the triple-layer stack (3 × (a)), the layers were deposited at 1500, 2000, and 1500 rpm, and for 3 × (b), at 1500, 2000, and 3000 rpm. The sixfold layer stack (6×) was spun at 1500, 2000, and 4 × 3000 rpm, while the ninefold layer stack (9×) was spun at 1500, 2000, and 7 × 3000 rpm.

Table 1.
The table presents the effective film thickness in nanometers (nm) for a multilayer m × PEDOT:PSS stack, with m being the number of PEDOT:PSS layers, 1, 3, 6, and 9.

One method of patterning PEDOT:PSS layers using sacrificial Ag layers has been described in our previous papers [39, 40]. The thickness of the PEDOT:PSS layer depends on the spinning speed, as shown in **Table 1**. For example, a triple PEDOT:PSS deposition at speeds of 1500, 2000, and 3000 rpm, for the 1st, 2nd, and 3rd deposition results in a total film thickness of 72 nm, which is significantly less than the additive sum of 65 + 56 + 29 = 150 nm and only 7 nm thicker than a single layer of PEDOT:PSS (65 nm) deposited at 1500 rpm. As mentioned in the introduction, the PEDOT and PSS components undergo vertical dephasing during spinning. The wet film consists of a PSS-rich solution at the top and conductive PEDOT strips at the bottom. These separated layers do not have sufficient time to blend during the subsequent soft bake at 120 degrees Celsius. The explanation for this phenomenon is provided in **Figure 2**, which schematically illustrates the PEDOT-rich bottom layer and the PSS-rich top layer. Naturally, there is a gradual transition between the two phases, with the PEDOT acting in the opposite direction at the bottom and a high PSS concentration at the top.

From our perspective, two factors contribute to the reduction in thickness observed during the deposition of multiple layers of PESOT:PSS. Firstly, the hydrophobic PEDOT-rich phase settles at the bottom of the PEDOT:PSS film during spin deposition, while the hydrophilic PSS-rich phase settles at the top. The hydrophilic PSS top part of the film is exposed to water during the second spin coating, and the PEDOT-rich phase partially replaces it. Thus, a significant reduction in layer thickness occurs during multiple depositions. This effect is prolonged for samples of 6× and 9 × PEDOT:PSS layers, as shown in **Table 1**. Therefore, upon subsequent PEDOT:PSS deposition, the PSS part is dissolved and removed, as well as some of the PSS inclusion within the bottom PEDOT layer. Eventually, in a multilayer PEDOT:PSS film, the top portion of the film stems from the last PEDOT:PSS deposition. The second factor is the densification of the PEDOT layer at the bottom. The removal of PSS inclusions creates more space to accommodate more PEDOT ribbons within essentially the same layer thickness.

With the use of the four-probe measuring approach, the sheet resistance (R_{sq}) and conductivity (σ) of the PEDOT:PSS layers have been found. For the sheet resistance measurements, we utilize the SP4 Four-Point Probe from Signatone 3014. This equipment features tungsten-carbide tips equally spaced at s = 0.0625″ between all four probes, aligned linearly. Each probe tip applies a pressure of 85 grams. The radius of the tungsten tip is 0.0016″. The Keithley 2400 serves as our source meter. Once configured for sheet resistance measurement, a constant fixed current is delivered to the outer two probes, while the inner two probes read voltages. The voltage difference between the inner two probes allows for the calculation of sheet resistance. Results are displayed on the computer's GUI. The currents supplied to the outer probes typically may range from 10 to 100 mA. Voltages are measured for each case, from which the sheet resistance value

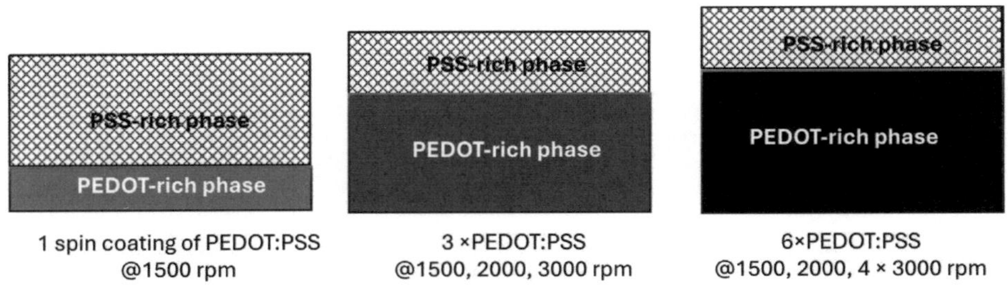

Figure 2.
Illustration of how the film's electrical conductivity is enhanced by multiple PEDOT:PSS coatings: Despite little difference in thickness between a single layer and multiple layers, the top PSS portion of the layer is gradually replaced with PEDOT ribbons, with the ultimate top layers consisting of PSS from the last PEDOT:PSS coating. Simultaneously, the PEDOT component is densified by subsequent coatings, as indicated in the **Figure 1** *by the increasing darkness of the fill.*

is internally calculated, yielding an output in Ohm/sq. The Keithley 2400 source meter is set to 4-wire CONFIG mode, evaluating sheet resistance using the formula R/sq. = $(\pi/\ln 2)\Delta V/I$. This formula applies when the spacings between neighboring probes exceed significantly the thickness of the film. Our PEDOT:PSS film thicknesses range from 30 to 213 nm, considerably smaller than the probe separation. The equipment's accuracy has been verified using standard samples of thin layers of known sheet resistance. Eq. (1) can be used to express the relationship between the engineering quantity of sheet resistance and the material quality of conductivity σ:

$$\sigma = \frac{1}{R_{sq} \times t} \tag{1}$$

Consider a pristine PEDOT:PSS film that was deposited at 2000 rpm, producing a sheet resistance of R_{sq}, or 1 MΩ/sq., with a film thickness (t) of 56 nm. Eq. (1) yields a conductivity (σ) value of 0.175 S/cm, which is equivalent to the value of sheet resistance.

In our recent efforts to improve the conductivity of PEDOT:PSS films through acid treatment methods, we performed a comparative analysis using three different acids: phosphoric, sulfuric, and nitric, with concentrations ranging from 0–100%. The duration of acid treatment varied from 1 to 60 seconds. One of our main studies was treating each acid at 100% concentration for a brief 2-second period, which significantly decreased the PEDOT:PSS films' sheet resistance. To identify the best acid type and concentration, we first treated a single layer of PEDOT:PSS. After precisely measuring the acid concentration, the PEDOT:PSS Si wafer samples were submerged in the acid solution for the allotted amount of time to begin the acid treatment process. To ensure that the results were accurate, the samples were promptly washed with DI water to get rid of any remaining acid. We then experimented with both topical and bulk doping techniques to apply silver nanoparticles (Ag NPs) to multilayer PEDOT:PSS films. To get a concentration of 0.5 mg/ml for bulk doping, we mixed 5 mg of Ag NPs with 10 ml of PEDOT:PSS solution. After at least an hour of stirring, this mixture was subjected to an additional hour of sonication. On the other hand, to prepare a dispersion of Ag NPs in ethanol for surface doping, we used the same 0.5 mg/ml concentration and the same stirring and sonication procedure. Next, at 1500 rpm, this dispersion was spin-coated onto the PEDOT:PSS layer's surface. Our earlier studies using copper nanoparticles [39, 40] led us to determine that the

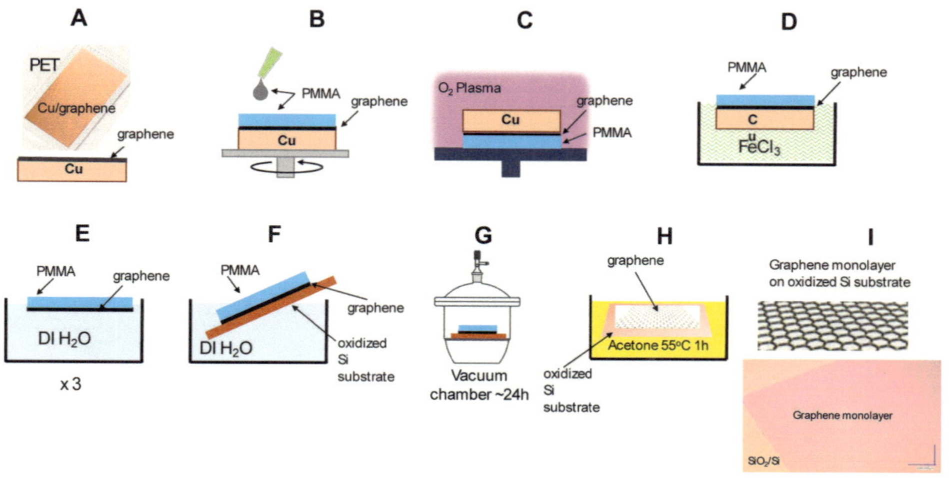

Figure 3.
Transfer process of graphene monolayer onto oxidized Si substrate: A. Cutting and taping polyethylene terephthalate (PET) underneath as-received graphene-on-copper sheets. B. Spin-coating of graphene with polyethylene methacrylate (PMMA) followed by an anneal at 50°C for 2 min. C. Removal of taped PET (not shown) and taping PMMA/Graphene/Cu sample upside down on to a substrate to etch backside graphene using oxygen plasma. D. Copper foil etched away in $FeCl_3$ solution until the entire copper sheet was dissolved. E. Transfer of PMMA with graphene using a glass slide into deionized (DI) water. F. Scooping of PMMA/Graphene using oxidized Si substrate. G. Storing PMMA/Graphene/SiO_2-Si under vacuum for at least 24 h to enhance graphene adhesion to the substrate. H. PMMA removed in warm (55°C) acetone for 1 hour. I. Final transferred Graphene monolayer on oxidized Si substrate after PMMA etch, cleaned with isopropyl alcohol (IPA) and dried, with optical microscope picture included.

concentration of silver nanoparticles at 0.5 mg/ml yielded the best outcomes. We employed the same concentration of silver nanoparticles to ensure uniformity. The silver nanoparticles were sourced from Sky Springs Nanomaterials, Inc., in two sizes: 20–30 nm and 50–60 nm. We observed that larger nanoparticles resulted in higher conductivity in PEDOT:PSS films, particularly with multiple coatings, as discussed in the following section.

Additionally, we present a novel method in this chapter for improving the performance of PEDOT:PSS films by sandwiching mono- and tri-layer graphene between the oxidized Si wafer and the films. As depicted in **Figure 3**, the transfer procedure for the graphene monolayer comprised multiple consecutive steps. The graphene monolayer was successfully transferred onto the wafer thanks to this process. Furthermore, Graphenea [60] supplied graphene tri-layers as separate samples on 1 cm × 1 cm pieces of oxidized Si wafers, obviating the requirement for further tri-layer graphene deposition on Si wafers.

The outcomes were then contrasted with the topical administration of copper nanoparticles (Cu NPs), a technique we had previously investigated [39, 40] because bulk doping with Cu NPs proved to be ineffective. Similar sheet resistance values were obtained by Ag and Cu NPs in the topical doping approach; these values ranged from 121 to 260 Ω/sq. for three layers of PEDOT:PSS to 173–256 Ω/sq., respectively. Additionally, both kinds of nanoparticles had comparable impacts on sheet resistance with three layers of PEDOT:PSS treated with 60% HNO_3 acid. Therefore, Cu nanoparticle topical doping is presented as a cost-effective alternative to silver nanoparticle doping. Both Ag and Au nanoparticles, when used in bulk doping, offer comparable conductivity improvements in PEDOT:PSS films, as documented in the literature [42, 43].

Sample	PEDOT:PSS Manufacturing Process	Sheet resistance [Ω/sq]
1	PEDOT:PSS 1 layer (2000 rpm)	1 M
2	PEDOT:PSS 1 layer (1500 rpm)	230 K
3	PEDOT:PSS 2 layers (3000 rpm)	85 k
4	PEDOT:PSS 2 layers (1500 rpm)	80 k
5	PEDOT:PSS 3 layers (1500 rpm)	60 k
6	PEDOT:PSS 3 layers (1500, 1500, 3000 rpm)	12 k
7	PEDOT:PSS 4 layers (1500, 2000, 3000, 3000 rpm)	4.8 k
8	PEDOT:PSS 5 layers (1500, 2000, 3000, 3000, 3000 rpm)	2.3 k
9	PEDOT:PSS 6 layers (1500, 2000, 4 × 3000 rpm)	0.98 k
10	PEDOT:PSS 9 layers (1500, 2000, 7 × 3000 rpm)	212
11	PEDOT:PSS 1 layer+Cu(25 nm)0.2 mg/ml 2000 rpm	1.2 k
12	PEDOT:PSS 2 layers+Cu(25 nm) 0.2 mg/ml @ 2000 rpm	467
13	PEDOT:PSS 3 layers (1500, 2000, 3000 rpm) + Cu(60 nm) 0.5 mg/ml	256
14	PEDOT:PSS 3 layers+Cu(60 nm) 0.2 mg/ml 1500 rpm	173
15	PEDOT:PSS 6 layers (1500, 2000, 4 × 3000 rpm) + Cu(60 nm) 0.2 mg/ml	62
16	PEDOT:PSS 1 layer (2000 rpm) + 60% HNO_3	182
17	PEDOT:PSS 3 layers (1500, 1500, 3000 rpm) + 60% HNO_3	51
18	PEDOT:PSS 6 layers (1500, 2000, 4 × 3000, rpm) + 60% HNO_3	22
19	PEDOT:PSS 9 layers (1500, 2000, 7 × 3000 rpm) + 60% HNO_3	17.5
20	n × PEDOT:PSS 6 layers (1500, 2000, 4 × 3000 rpm) + 60% HNO_3 each layer treated	10.2
21	n × PEDOT:PSS 6 layers (1500, 2000, 4 × 3000 rpm) + 60% HNO_3 only top layer treated	7.7
22	PEDOT:PSS 9 layers (1500, 2000, 7 × 3000 rpm) + graphene monolayer	6.7

The steps of the applied electrical enhancement method are covered in the book.

Table 2.
Samples of PEDOT:PSS film(s) with different processing stages and the resulting sheet resistance.

3. Analysis of the resulting conductivity enhancements

The key results of this study are summarized in **Table 2**, which presents the sheet resistance measurements for both single and multiple layers of PEDOT:PSS, with and without acid treatment, and with and without the incorporation of Cu nanoparticles. Separate Tables provide details on samples treated with bulk and topical Ag NP doping, as well as the sheet resistance variations based on different acid concentrations and treatment durations. One important observation is that spinning speed greatly influences sheet resistance. For instance, when a single layer of PEDOT:PSS is spun at 2000 rpm, sample 1 displays a sheet resistance of 1 MΩ, but sample 2 exhibits a resistance of 230 kΩ at 1500 rpm. Moreover, the sheet resistance decreases with increasing layer count: sample 2's 230 kΩ decreases to sample 3's 85 kΩ, sample 5's 60 kΩ, sample 7's 4.8 kΩ, sample 8's 2.3 kΩ, sample 9's 0.98 kΩ, and

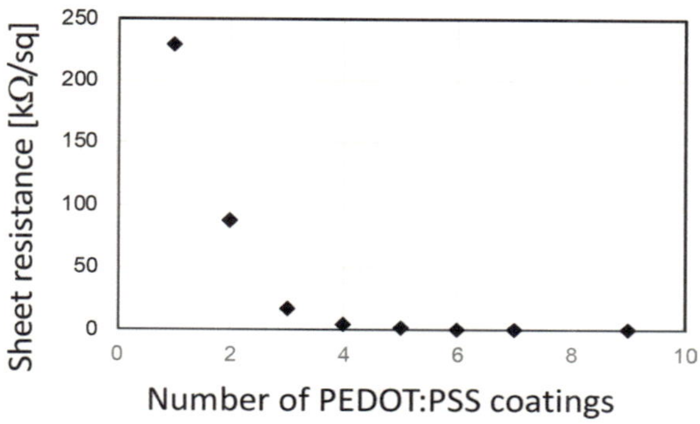

Figure 4.
The sheet resistance of n-layers of PEDOT:PSS layers as a function of the number of layers. The number n of PEDOT:PSS layers ranges from 1 to 9.

sample 10's 212 Ω; these correspond to 2, 3, 4, 5, 6, and 9 layers of PEDOT:PSS, respectively. **Figure 4** illustrates this pattern of diminishing sheet resistance with more PEDOT:PSS layers. As the number of layers grows, an exponential drop in sheet resistance is observed. The following formula can be used to approximate the overall trend of sheet resistance:

$$R_{sp}(n) = 827k\Omega \times \exp\left(-\frac{n}{0.78}\right) \qquad (2)$$

In **Figure 5**, AFM images depicting the surface roughness of the multi-stack PEDOT:PSS films are presented. It is evident that the surface roughness for 3×, 6×,

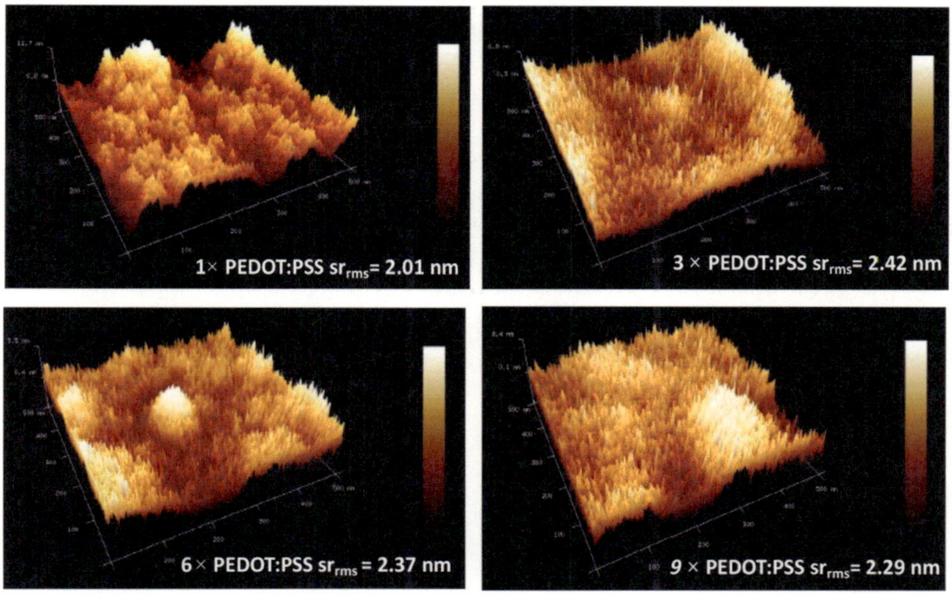

Figure 5.
Surface roughness pictures by AFM of a single, triple, sixfold, and ninefold PEDOT:PSS layer stacks. The rms values of the surface roughness are also indicated.

Acid	No acid	HNO_3	H_3PO_4	H_2SO_4
Sheet Res. [Ω/sq]	18 k	51	65	356

There was a 100% concentration of every acid.

Table 3.
Effects of HNO_3, H_3PO_4, and H_2SO_4 treatments on the sheet resistance, Rsq of a single PEDOT:PSS layer

and 9 × PEDOT:PSS layers is only marginally higher than that observed for a single PEDOT:PSS layer, and when the number of PEDOT:PSS layers increases, there seems to be a minor reduction in roughness. Three acids were used to treat the multilayer PEDOT:PSS stack: HNO_3, H_2SO_4, and H_3PO_3. We discovered that, out of these three acids, HNO_3 acid significantly increased conductivity. **Table 3** shows how the three acids, applied for two seconds to a single layer of PEDOT:PSS at 100% concentration, reduced sheet resistance.

In additional multilayer PEDOT:PSS stacks, with acid concentrations ranging from 20–100% and treatment times ranging from two seconds to one-minute, comparable decreases in sheet resistance have been noted. Nitric acid proved to be the most effective, so further comparisons focus on HNO_3 specifically. The treatment process for a multilayer PEDOT:PSS stack is depicted in **Figure 6**. As additional layers of PEDOT:PSS are applied, the PEDOT phase tends to migrate downward to the bottom of the layer, while the water-soluble PSS phase forms the upper portion of each layer. When a new layer is deposited, the PEDOT component of this layer penetrates the soluble PSS of the previous one, leading to a thicker and more compact PEDOT phase at the bottom. Upon administering acid treatment, the upper nonconductive PSS segment is removed. Additionally, a significant amount of PSS inclusions within the PEDOT phase is likely removed during acid treatment, which enhances the conductivity of the final layer. **Figure 7** illustrates the impact of HNO_3 treatment on the multilayer stack, displaying the sheet resistance on a logarithmic scale relative to the number of PEDOT:PSS coatings. This data is compared to untreated samples, as shown in **Figure 4**. The results clearly demonstrate that acid treatment consistently reduces sheet resistance across all multilayer stacks.

We also investigated the impact of acid concentration on a single PEDOT:PSS layer's sheet resistance. The results for a single PEDOT:PSS layer treated with HNO_3 acid at five different doses for 30, 60, and 90 seconds are shown in **Table 4**.

Table 4 indicates that a 40% or greater acid content is needed to significantly lower the sheet resistance. The changes in reduction factors throughout acid

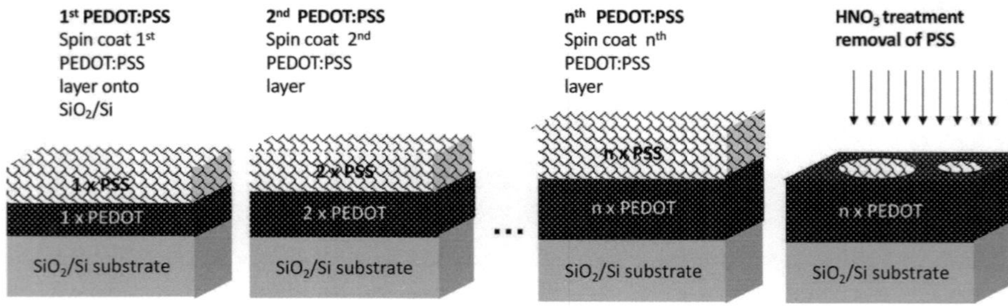

Figure 6.
Process sequence for treating multiple PEDOT:PSS layers with HNO_3 acid involves two methods. In one method, illustrated in the figure, acid treatment occurs after the deposition of the final PEDOT:PSS coating. In the other method, each PEDOT:PSS layer is exposed to acid treatment individually.

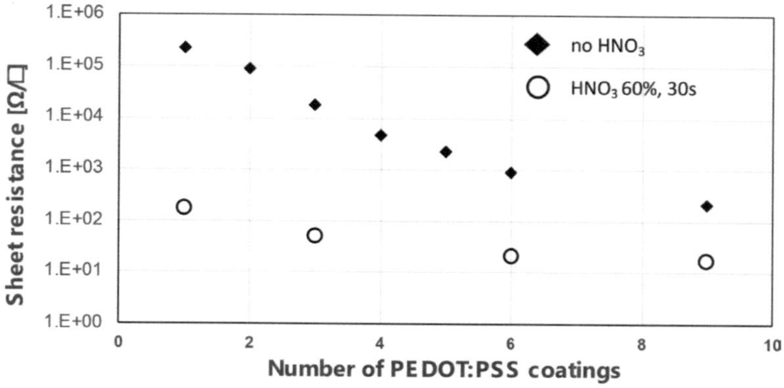

Figure 7.
Sheet resistance, depicted on a logarithmic scale, of multilayer PEDOT:PSS as a function of the number of layers, with and without HNO_3 acid treatment. The acid treatment has been applied after the deposition of the respective final layer.

HNO_3 conc.	100%	80%	60%	40%	20%
R_{sq} [Ω/sq] for 30 s	206	250	240	290	10,830
R_{sq} [Ω/sq] for 60 s	215	251	255	280	7823
R_{sq} [Ω/sq] for 90 s	228	245	193	270	5488

About 100% concentration in the instance of HNO_3 is equivalent to the commercially available 14 M HNO_3.

Table 4.
Effects of HNO_3 treatment at different concentration levels on a sheet resistance (R_{sq}) of a single-layer PEDOT:PSS for three different acid treatment times (30, 60, and 90 seconds).

concentrations are generally minor, spanning between 130 and 201 Ω/sq.; however, the largest reduction in R_{sq} is recorded at the highest concentration of 100%.

We also investigated the long-term stability of the acid treatment. Samples were left to age for several days, during which the sheet resistance was periodically measured. A slight increase in sheet resistance was noted over the initial days, with the most pronounced change occurring within the first 24 hours. After approximately 3–4 days, this increase stabilized to a steady value. The least degradation was observed in PEDOT:PSS treated with 20% HNO_3, while the greatest degradation occurred with 100% HNO_3, as shown in **Table 5**. The degradation of sheet resistance is calculated using the formula $DEG = (R_{sq}(m = 0)-R_{sq}(m))/R_{sq}(m = 0)$, where $R_{sq}(m)$ represents the sheet resistance measured on the m_{th} day after production.

In **Table 6** the sheet resistance of 1 × PEDOT:PSS layer treated in 100% nitric acid for three durations of 1, 2, and 2 s, is shown. While the sheet resistance does not change much in the initial phase of acid treatment, there is a slight increase of sheet resistance at a treatment with 100% acid concentration for 3 s.

HNO_3 conc.	100%	80%	60%	40%	20%
DEG [%] after 8 days	160	138	110	103	82

Table 5.
Degradation of R_{sq} for a single layer of PEDOT:PSS treated with nitric acid at different concentration levels during an eight-day exposure to laboratory ambient conditions.

Duration HNO_3 100% [s]	**1**	**2**	**3**
Sheet Res. [Ω/sq]	60	60	65

Table 6.
Effect of treatment with HNO_3 at 100% concentration on the sheet resistance (R_{sq}) of a single-layer PEDOT:PSS for three acid treatment durations: 1 s, 2 s, and 3 s, probably due to some removal of PEDOT fibers.

The sheet resistance results indicate two competing effects: the acid treatment predominantly eliminates the insulating PSS phase, thereby reducing the film's overall thickness. However, it may also remove portions of the conductive PEDOT phase, potentially increasing the sheet resistance.

Samples 16 to 19 in **Table 2** show the effect of HNO_3 treatment, on 1×, 3×, 6×, 9× multiple PEDOT:PSS layer stack. The sheet resistance decreases from 182 Ω/sq. for a single PEDOT:PSS layer to 17.5 Ω/sq. for a 9-layer stack. Multilayer PEDOT:PSS coatings underwent a single HNO_3 treatment, which took place following the final PEDOT:PSS coating's deposition. This begged the question of what would happen if the HNO_3 treatment was administered after each PEDOT:PSS coating separately. To determine the efficacy of the HNO_3 treatment for a multilayer PEDOT:PSS stack, we compared two 6-layer PEDOT:PSS stacks: one stack was manufactured by depositing six PEDOT:PSS layers consecutively, with the final layer treated by HNO_3 (sample 20 in **Table 2**); the other stack was produced by depositing six layers of PEDOT:PSS, with each PEDOT:PSS layer treated by HNO_3 before the next layer was applied (sample 21 in **Table 2**). The resulting sheet resistance of the latter sample was 10.2 Ω (sample 20 in **Table 2**). This can be compared to the sheet resistance of 7.7 Ω/sq. for the sample with 6 × PEDOT:PSS coatings treated with HNO_3 once upon the application of the sixth PEDOT:PSS coating (sample 21 in **Table 2**). It is observed that the treatment of each PEDOT:PSS layer with HNO_3 provides a slightly higher sheet resistance than a single HNO_3 treatment of the 6-layer stack. This suggests again that while HNO_3 exposure predominantly removes the PSS top layer and PSS inclusions within the PEDOT phase, some portion of PEDOT may also be removed.

We now shift our attention to investigating the possibility of using copper and silver nanoparticles (Cu, Ag NP) to improve PEDOT:PSS electrical conductivity. Samples 11 to 15 show the effect of a topical doping with Cu NP, on 1×, 3×, 6× multiple PEDOT:PSS layer stack. The sheet resistance decreases from 1.2 kΩ/sq. for a single PEDOT:PSS layer to 62 Ω/sq. for a 6-layer stack. Because silver is inert, we focused on employing Ag nanoparticles (Ag NP). To that end, we used two introduction techniques: topical administration, which we used after the PEDOT:PSS layer had been soft baked, and bulk doping, in which Ag NP was added directly into the PEDOT:PSS solution. Ag nanoparticles with two distinct sizes—20–30 and 50–60 nm—were used. Larger Ag nanoparticles consistently result in better conductivity, as demonstrated by our findings. The conductivity values for bulk and topical applications of Ag NP in 1 × PEDOT:PSS and 3 × PEDOT:PSS combinations are compared in **Table** 7.

Topical application of Ag NP leads to a significantly lower sheet resistance compared to the bulk method, showcasing a three-order-of-magnitude enhancement. Therefore, this approach might be preferred in applications such as ReRAM cells where the presence of Cu or Ag is highly advantageous [61–65].

These findings can be contrasted with PEDOT:PSS layers doped with Cu nanoparticles instead of Ag nanoparticles. Notably, only topical doping was employed for Cu NP (samples 11–15 in **Table 2**) as bulk doping with Cu does not improve the sheet

Ag 0.5 mg/ml NP 50–60 nm	Topical doping 1 × PEDOT:PSS	Bulk Doping 1 × PEDOT:PSS	Topical doping 3 × PEDOT:PSS	bulk doping 3 × PEDOT:PSS
R_{sq} [Ω/sq]	464	629 k	121	30 k

Table 7.
Effect of topical and bulk doping using 50–60 nm Ag NP at 0.5 mg/ml on the sheet resistance (R_{sq}) of a triple PEDOT:PSS layer stack (3 × PEDOT:PSS) and a PEDOT:PSS monolayer (1 × PEDOT:PSS).

Cu NP 0.5 mg/ml NP 60 nm	Topical doping 1 × PEDOT:PSS	Topical Doping 2 × PEDOT:PSS	Topical doping 3× PEDOT:PSS
R_{sq} [Ω/sq]	756	467	173

The value for 3× PEDOT:PSS is copied from **Table 2** *(sample 14).*

Table 8.
The effects of topical doping with 60 nm-sized Cu nanoparticles (Cu NP) at a concentration of 0.5 mg/ml on sheet resistance (R_{sq}) were studied across different PEDOT:PSS layer configurations: a single monolayer (1 × PEDOT:PSS), a double layer (2 × PEDOT:PSS), and a triple layer (3 × PEDOT:PSS).

resistance due to Cu oxidation in the aqueous PEDOT:PSS dispersion. **Table 8** displays the corresponding sheet resistances for Cu NP doping.

The results presented in **Tables 7** and **8** indicate that bulk doping is not effective even with noble metal nanoparticles. This is probably because of issues such as poor Ag NP integration inside PEDOT:PSS and the potential for Ag NP to be coated by PSS. The data from **Tables 7** and **8** when compared show that topical doping—with or without Ag—performs better than bulk doping. Notably, topical doping with Cu NP has efficacy comparable to that of Ag NP doping. For three PEDOT:PSS layers, Ag and Cu NP produced comparable sheet resistance values of 121 and 184 Ω/sq., respectively. Furthermore, when these layers were treated with 60% HNO_3 acid, both nanoparticles had similar effects on sheet resistance. Although Ag NP offered slightly better results than Cu NP, Cu is more cost-effective, making it the preferred doping option for PEDOT:PSS films.

This suggests that Cu nanoparticles present an economically viable alternative to silver particles for doping PEDOT:PSS films. The literature also reports that bulk doping with Ag and Au nanoparticles provides similar improvements in conductivity for PEDOT:PSS films, reinforcing our earlier conclusion that topical doping is more effective than bulk doping, likely applying to Au NP as well.

To lessen sheet resistance, another technique was to place graphene on an oxidized Si wafer before adding the PEDOT:PSS layers. Two types of graphene were investigated, as was previously mentioned: a large cm-sized single-layer graphene monolayer and a tri-layer graphene. The tri-layer displayed a substantially lower resistance of 50 Ω/sq. than the monolayer alone, which had a sheet resistance of 20 kΩ/sq. However, utilizing tri-layer graphene resulted in adhesion problems with PEDOT:PSS. We treated the tri-layer graphene with plasma to address this issue, which greatly enhanced PEDOT:PSS adhesion. Unfortunately, this enhancement also caused a notable decrease in the tri-layer graphene's sheet resistance. Additionally, the R_{sq} values varied widely from 14 Ω/sq. to 4 kΩ/sq., indicating a highly nonuniform distribution of sheet resistance. This substantial nonuniformity led us to largely abandon further attempts with tri-layer graphene.

Despite this, we successfully synthesized a single sample of 6 × PEDOT:PSS on plasma-treated tri-layer graphene, achieving a sheet resistance of 1350 Ω/sq. One positive outcome was the uniformity of sheet resistance across the wafer. On the other hand, when paired with 9 × PEDOT:PSS layers, a graphene monolayer yielded the lowest sheet resistance of 6.7 Ω/sq. (sample 22 in **Table 2**). Interestingly, the sheet resistance of 9 × PEDOT:PSS treated with 60% nitric acid, resulting in R_{sq} = 7.7 Ω/sq. (sample 21 in **Table 2**), closely approaches this outcome. Notably, this method is significantly simpler and more cost-effective in terms of manufacturing.

Significantly, achieving sheet resistance values as low as 14 Ω/sq. at the lower end of the spectrum through plasma treatment of tri-layer graphene offers promising prospects for future exploration. Given the relative simplicity of dealing with tri-layer graphene as opposed to the difficult process of depositing a graphene monolayer onto a substrate, there is a real chance to address the problem of non-uniformity. However, it is worth noting that tri-layer graphene necessitates a film provided on oxidized Si substrates by the supplier (Graphenea [60]), which limits autonomy in the manufacturing process on real-life wafer sizes.

Let us now examine the combined effects of various enhancement techniques on PEDOT:PSS conductivity. To our knowledge, the synergistic effects of multiple enhancement methods have not been documented. We first explored the combination of silver nanoparticle doping with nitric acid treatment. On its own, HNO_3 treatment of 3 × PEDOT:PSS results in a sheet resistance of 50 Ω/sq., and similarly, applying Ag NP topically to the PEDOT:PSS stack achieves a comparable outcome. This suggests that adding metal nanoparticles to the acid treatment does not offer significant additional benefits.

Another interesting combination is nitric acid treatment paired with the insertion of a graphene monolayer. While both methods independently reduce sheet resistance considerably, their combined effect is minimal. For instance, treating a 9 × PEDOT:PSS layer with nitric acid reduces sheet resistance to 7.7 Ω/sq., and adding a graphene monolayer only further lowers it slightly to 6.7 Ω/sq.

As a result of this research, the best way to increase PEDOT:PSS conductivity is to combine a multilayer coating technique with nitric acid treatment. Other enhancing technique combinations do not offer any more advantages. Hence, we wish to emphasize that the nitric acid approach is the most economical and simple choice when considering graphene insertion and metal nanoparticle doping.

4. Conclusions

We have expanded upon our previous research [39, 40], with the aim of mitigating the sheet resistance of a PEDOT:PSS film through novel techniques to enhance its electrical conductivity. These techniques include topical and bulk doping with Ag and Cu metal nanoparticles, as well as acid treatment with three distinct acids (HNO_3, H_2SO_4, and H_3PO_4) at varied doses and treatment periods. We also looked at the application of graphene in two different forms: a single layer and a tri-layer. Our investigation concentrated on how various enhancing methods, taken separately and in combination, reduce sheet resistance. The results show that whereas most combination approaches only yield marginal benefits that frequently do not justify their cost and practicality, adding graphene, as either a monolayer or tri-layer, dramatically lowers PEDOT:PSS's sheet resistance. The combination of nitric acid treatment with multiple PEDOT:PSS coatings is the most cost-effective and technologically simple strategy explored, when compared to alternatives like graphene insertion and metal nanoparticle doping. Following closely behind is the effectiveness of a combination involving multiple PEDOT:PSS coatings paired with topical doping by Cu NP. We anticipate that the insights gleaned from our exploration of multiple conductivity enhancement techniques presented in this chapter will assist fellow researchers in determining the most suitable approaches for enhancing conductivity of organic polymers within their respective fields of application.

Author details

Aaron DiFilippo, Amrita Chakraborty and Marius K. Orlowski*
Bradley Department of Computer and Electrical Engineering, Virginia Tech, Blacksburg, USA

*Address all correspondence to: m.orlowski@vt.edu

References

[1] Ahmad Shahrim NA, Ahmad Z, Azman AW, Buysc YF, Sarifuddina N. Mechanisms for doped PEDOT:PSS electrical conductivity improvement. Materials Advances. 2021;**2**:7118-7138

[2] Mahato S, Gerling LG, Voz C, Alcubilla R, Puigdollers J. PEDOT:PSS as an alternative hole selective contact for ITO-free hybrid crystalline silicon solar cell. IEEE Journal of Photovoltaics. 2016;**6**(4):934

[3] Alemu D, Wei HY, Ho KC, Chu CW. Highly conductive PEDOT:PSS electrode by simple film treatment with methanol for ITO-free polymer solar cells. Energy & Environmental Science. 2012;**5**:9662-9671

[4] Ouyang J, Chu CW, Chen FC, Xu Q, Yang Y. High-conductivity poly(3,4-ethylenedioxythiophene):Poly(styr ene sulfonate) film and its application in polymer optoelectronic devices. Advanced Functional Materials. 2005;**15**:203-208

[5] Hsiao YS, Whang WT, Chen CP, Chen YC. High-conductivity poly(3,4-ethylenedioxythiophene):poly(styr ene sulfonate) film for use in ITO-free polymer solar cells. Journal of Materials Chemistry. 2008;**18**:5948-5955

[6] Kim N, Kee S, Lee SH, Lee BH, Kahng YH, Jo Y-R, et al. Highly conductive PEDOT:PSS nanofibrils induced by solution-processed crystallization. Advanced Materials. 2014;**26**:2268-2272

[7] Kim S, Sanyoto B, Park WT, Kim S, Mandal S, Lim JC, et al. Purification of PEDOT:PSS by ultrafiltration for highly conductive transparent electrode of all-printed organic devices. Advanced Materials. 2016;**28**:10149-10154

[8] Jang H, Kim MS, Jang W, Son H, Wang DH, Kim FS. Highly conductive PEDOT:PSS electrode obtained via post-treatment with alcoholic solvent for ITO-free organic solar cells. Journal of Industrial and Engineering Chemistry. 2020;**86**:205-210

[9] Jeong SH, Ahn S, Lee TW. Strategies to improve electrical and electronic properties of PEDOT:PSS for organic and perovskite optoelectronic devices. Macromolecular Research. 2019;**27**:2-9

[10] Groenendaal L, Jonas F, Freitag D, Pielartzik H, Reynolds JR. Poly(3,4-ethylenedioxythiophene) and its derivatives: Past, present, and future. Advanced Materials. 2000;**12**:481-494

[11] Kirchmeyer S, Reuter K. Scientific importance, properties and growing applications of poly(3,4-ethylenedioxythiophene). Journal of Materials Chemistry. 2005;**15**:2077-2088

[12] Elschner A, Lövenich W. Solution-deposited PEDOT for transparent conductive applications. MRS Bulletin. 2011;**36**:794-798

[13] Hecht DS, Hu L, Irvin G. Emerging transparent electrodes based on thin films of carbon nanotubes, graphene, and metallic nanostructures. Advanced Materials. 2011;**23**:1482-1513

[14] Na SI, Kim SS, Jo J, Kim DY. Efficient and flexible ITO-free organicosolar cells using highly conductive polymer anodes. Advanced Materials. 2008;**20**:2061-2067

[15] Ramizy A, Hassan Z, Omar K, Al-Douri Y. New optical features to enhance solar cell performance based on porous silicon surfaces. Applied Surface Science. 2011;**257**:6112-6117

[16] Earmme T, Jenekhe SA. Solution-processed, alkali metal-salt-doped, electron-transport layers for high-performance phosphorescent organic light-emitting diodes. Advanced Functional Materials. 2012;**22**:5126-5136

[17] Cha H, Li J, Li Y, Kim SO, Kim YH, Kwon SK. Effects of bulk heterojunction morphology control via thermal annealing on the fill factor of anthracene-based polymer solar cells. Macromolecular Research. 2020;**28**:820-825

[18] Wang L, Park JS, Lee HG, Kim GU, Kim D, Kim C, et al. Impact of chlorination patterns of naphthalenediimide-based polymers on aggregated structure, crystallinity, and device performance of all-polymer solar cells and organic transistors. ACS Applied Materials & Interfaces. 2020;**12**:56240-56250

[19] Österholm AM, Shen DE, Dyer AL, Reynolds JR. Optimization of PEDOT films in ionic liquid supercapacitors: Demonstration as a power source for polymer electrochromic devices. ACS Applied Materials & Interfaces. 2013;**5**:13432-13440

[20] Gulercan D, Commandeur D, Chen Q, Sarac AS. A ternary PEDOT-TiO_2-reduced graphene oxide nanocomposite for supercapacitor applications. Macromolecular Research. 2019;**27**:867-875

[21] Bernards DA, Malliaras GG. Steady-state and transient behavior of organic electrochemical transistors. Advanced Functional Materials. 2007;**17**:3538-3544

[22] Rivnay J, Leleux P, Ferro M, Sessolo M, Williamson A, Koutsouras DA, et al. High-performance transistors for bioelectronics through tuning of channel thickness. Science Advances. 2015;**1**:e1400251

[23] Harman DG, Gorkin IR, Stevens L, Thompson B, Wagner K, Weng B, et al. Poly(3,4-ethylenedioxythiophene): Dextran sulfate (PEDOT:DS)–a highly processable conductive organic biopolymer. Acta Biomaterialia. 2015;**14**:33-42

[24] Chou TR, Chen SH, Chiang YT, Lina YT, Chao CY. Highly conductive PEDOT:PSS films by posttreatment with dimethyl sulfoxide for ITO-free liquid crystal display. Journal of Materials Chemistry C. 2015;**3**:3760

[25] Sinh AK, Dwivedi ADD, Pandey A, Chakrabarti P. Design and implementation of an inverter and its application in ring oscillator circuits using an organic-thin-film-transistor based on an FTM-derived channel. Semiconductor Science and Technology. 2021;**36**:125006. DOI: 10.1088/1361-6641/ac274f

[26] Dadhich S, Mathur G, Dwivedi ADD. Mobile charge carrier based modeling of 4H–21 DNTT and structure analysis of OTFT. In: 2021 IEEE 2nd International Conference on Technology, Engineering, Management for Societal Impact Using Marketing, Entrepreneurship and Talent (TEMSMET). Pune, India: Publisher IEEE; Dec 2021. pp. 1-5. DOI: 10.1109/TEMSMET53515.2021.9768714

[27] Nardes AM, Kemerink KJ, Janssen RA, Bastiaansen AM, Kiggen NMM, Langeveld BMW, et al. Microscopic understanding of the anisotropic conductivity of PEDOT:PSS thin films. Advanced Materials. 2007;**19**:1196-1200. DOI: 10.1002/adma.200602575

[28] Zabihi F, Xie Y, Gao S, Eslamian M. Morphology, conductivity, and wetting

characteristics of PEDOT:PSSthin films deposited by spin and spray coating. Applied Surface Science. 2015;**338**:163-177

[29] Menard E, Meitl MA, Sun YG, Park JU, Shir DJL, Nam YS, et al. Micro- and nanopatterning techniques for organic electronic and optoelectronic systems. Chemical Reviews. 2007;**107**:1117

[30] Sheats JR. Manufacturing and commercialization issues in organic electronics. Journal of Materials Research. 2004;**19**:1974-1989. DOI: 10.1557/JMR.2004.0275

[31] de Gans BJ, Duineveld PC, Schubert US. Inkjet printing of polymers: State of the art and future developments. Advanced Materials. 2004;**16**:203

[32] Ling MM, Bao ZN. Thin film deposition, patterning, and printing in organic thin film transistors. Chemistry of Materials. 2004;**16**:4824

[33] Stewart MD, Johnson SC, Sreenivasan SV, Resnick DJ, Willson CG. Nanofabrication with step and flash imprint lithography. Journal of Microlithography, Microfabrication, and Microsystems. 2005;**4**:011002

[34] Khang DY, Kang H, Kim T, Lee HH. Low-pressure nanoimprint lithography. Nano Letters. 2004;**4**:633

[35] Huang J, Xia R, Kim Y, Wang X, Dane J, Hofmann O, et al. Patterning of organic devices by interlayer lithography. Journal of Materials Chemistry. 2007;**17**:1043

[36] Ouyang S, Xie Y, Wang D, Zhu D, Xu X, Tan T, et al. Surface patterning of PEDOT:PSS by photolithography for organic electronic devices. Hindawi Publ. Corp. Journal of Nanomaterials. 2015. article ID 603148. DOI: 10.1155/2015/603148

[37] Taylor PG, Lee JK, Zakhidov AA, Chatzichristidi M, Fong HH, DeFranco JA, et al. Orthogonal patterning of PEDOT:PSS for organic electronics using hydrofluoroether solvents. Advanced Materials. 2009;**21**:2314-2317

[38] Son D, Kim JH, Park D, Choi WK, Li F, Hun Ham J, et al. Nanovolatile flexible organic bistable devices fabricated utilizing CdSe/ZnS nanoparticles embedded in a conducting poly N-vinylcarbazole polymer layer. Nanotechnology. 2008;**19**:055204

[39] Chakraborty A, Herrera D, Fallen P, Bampton N, Hall D, Olivero T, et al. Conductive organic electrodes for flexible electronic devices. Scientific Reports. 2023;**13**:412

[40] Chakraborty A, DiFilippo A, Deivasigamani S, Hong C, Madwesh A, Orlowski M. Exploring novel methods: Acid treatment, metal nanoparticle doping, and graphene insertion for enhanced electrical conductivity of nm thin PEDOT:PSS films. Synthetic Metals. 2024;**307**:117694

[41] Zhang X, Wu J, Wang J, Zhang J, Yang Q, Fu F, et al. Highly conductive PEDOT:PSS transparent electrode prepared by a post-spin-rinsing method for efficient ITO-free polymer solar cells. Solar Energy Materials and Solar Cells. 2016;**144**:143-149

[42] Patil DS, Pawar SA, Hwang J, Kim JH, Patil PS, Shin JC. Silver incorporated PEDOT:PSS for enhanced electrochemical performance. Journal of Industrial and Engineering Chemistry. 2016;**42**:113-120

[43] Zhang RC, Sun DR, Zhang R, Lin WF, Macias-Montero M, Patel J,

et al. Gold nanoparticle-polymer nanocomposites synthesized by room temperature plasma and their potential for fuel cell electrocatalytic application. Scientific Reports. 2017;7:46682

[44] Ghazy OA, Ibrahim MM, Abou-Elfadl FI, Hosni HM, Shehata ME, Dehiedy NM, et al. PEDOT:PSS incorporated silver nanoparticles prepared by gamma radiation for the application in organic solar cells. Journal of Radiation Research and Applied Sciences. 2015;**8**:166-172

[45] Pham LQ, Sohn JH, Kim CW, Park JH, Kang HS, Lee BC, et al. Copper nanoparticles incorporated with conducting polymer: Effects of copper concentration and surfactants on the stability and conductivity. Journal of Colloid and Interface Science. 2012;**365**:103-109

[46] Meng W, Ge R, Li Z, Tong J, Liu T, Zhao Q, et al. Conductivity enhancement of PEDOT:PSS films via phosphoric acid treatment for flexible all-plastic solar cells. ACS Applied Materials & Interfaces. 2015;7:14089-14094

[47] Hu Z, Wang J, Wang Z, Gao W, An Q, Zhang M, et al. Semitransparent ternary nonfullerene polymer solar cells exhibiting 9.40% efficiency and 24.6% average visible transmittance. Nano Energy. 2019;**55**:424-432

[48] Hu Z, Wang Z, An Q, Zhang F. Semitransparent polymer solar cells with 12.37% efficiency and 18.6% average visible transmittance. Science Bulletin. 2020;**65**(2):131-137

[49] Yoo D, Kim J, Kim JH. Direct synthesis of highly conductive PEDOT:PSS/graphene composites and their applications in energy harvesting systems. Nano Research. 2014;7:717-730

[50] Zhang M, Yuan W, Yao B, Li C, Shi G. Solution-processed PEDOT:PSS/graphene composites as the electrocatalyst for oxygen reduction reaction. ACS Applied Materials & Interfaces. 2014;**6**:3587-3593

[51] Mahakul PC, Sa K, Das B, Subramaniam BV, Saha S, Moharana B, et al. Preparation and characterization of PEDOT:PSS/reduced graphene oxide-carbon nanotubes hybrid composites for transparent electrode applications. Journal of Materials Science. 2017;**52**:5696-5707

[52] Kepic DP, Markovic ZM, Jovanovic SP, Peruško DB, Budimir MD, Holclajtner-Antunovic ID, et al. Preparation of PEDOT:PSS thin films doped with graphene and graphene quantum dots. Synthetic Metals. 2014;**198**:150-154

[53] Liu D, Rahman MM, Ge C, Kima J, Lee JJ. Highly stable and conductive PEDOT:PSS/graphene nanocomposites for biosensor applications in aqueous medium. New Journal of Chemistry. 2017;**41**:15458-15465

[54] Ko SH, Kim SW, Lee YJ. Flexible sensor with electrophoretic polymerized graphene oxide/PEDOT:PSS composite for voltametric determination of dopamine concentration. Scientific Reports. 2021;**11**:21101

[55] Du FP, Cao NN, Zhang YF, Fu P, Wu YG, Lin ZD, et al. PEDOT:PSS/graphene quantum dots with enhanced thermoelectric properties via strong interfacial interactions and phase separation. Scientific Reports. 2018;**8**:6441

[56] Badri MAS, Noor NF, Zain AR, MatSalleh M, Tengku Aziz TH. Exfoliated graphene-alkaline lignin-PEDOT:PSS composite as a transparent

conductive electrode. Nanomaterials and Nanotechnology. 2021;**11**:1-8

[57] Nemani SK, Chen D, Mohamed MH, Sojoudi H. Stretchable and hydrophobic electrochromic devices using wrinkled graphene and PEDOT:PSS. Journal of Nanomaterials. 2018;**4**:3230293

[58] Adekoya GJ, Sadiku RE, Ray SS. Nanocomposites of PEDOT:PSS with graphene and its derivatives for flexible electronic applications: A review. Macromolecular Materials and Engineering. 2021;**306**:2000716

[59] Faruk O, Adak B. Recent advances in PEDOT:PSS integrated graphene and MXene-based composites for electrochemical supercapacitor applications. Synthetic Metals. 2023;**297**:117384

[60] Customer sends sends his substrate to graphenea laboratory for a transfer of the graphene monolayer on top of it. Examples of substrates successfully transferred graphene to: devices, solar cells, LEDs, gold, alumina, silver mirrors, sapphire. Available from: https://www.graphenea.com/products/graphene-on-your-substrate

[61] Liu T, Kang Y, Verma M, Orlowski M. Switching characteristics of antiparallel resistive switches. IEEE Electron Device Letters. 2012;**33**:429

[62] Kang Y, Ruan H, Claus RO, Heremans J, Orlowski M. Observation of quantized and partial quantized conductance in polymer-suspended graphene nanoplatelets. Nanoscale Research Letters. 2016;**11**:179

[63] Liu T, Kang Y, El-Helw S, Potnis T, Orlowski M. Physics of the voltage constant in multilevel switching of conductive bridge resistive memory. Japanese Journal of Applied Physics. 2013;**52**:084202

[64] Al-Mamun MS, Chakraborty A, Orlowski M. Analysis of the electrical ReRAM device degradation induced by thermal crosstalk. Advanced Electronic Materials. 2023;**9**(4):2201081

[65] Chakraborty A, Al-Mamun MS, Orlowski MK. In: Chang Y-F, editor. Thermal Reliability Issues in ReRAM Memory Arrays Chapter in the Book: Y.- F. Chang, Ed., 'Memristors - The Fourth Fundamental Circuit Element - Theory, Device, and Applications'. IntechOpen. 07 Feb 2024. DOI: 10.5772/intechopen.110931

Section 2

Organic Electronics: Device Applications

Chapter 4

Numerical Simulation and Physics-Based Modelling of Ph-BTBT-C10-Based Organic Thin Film Transistor

Shubham Dadhich, Arun Dev Dhar Dwivedi and Ram Babu Pareek

Abstract

Organic thin film transistors (OTFTs) are an important aspect in the development of low-cost, large-area, and flexible ICs, underscoring the necessity for precise modelling and understanding of device physics. A finite element method (FEM)-based numerical device simulation of OTFTs is presented in this chapter using Ph-BTBT-C10 as the semiconductor. This model includes field-dependent mobility and a double-peak density of states. This model extends existing theories by integrating temperature-dependent band gap narrowing with the carrier concentration effects caused by heavy doping. It is more accurate in predicting the band gap under different thermal conditions and doping concentrations, which is essential for fabricating advanced semiconductor devices. For charge-carrier movements, hopping mobility describes the transport of charge through localised states in semiconductors influenced by thermal excitations but Poole-Frenkel mobility describes a field-induced charge movement, particularly at high electric fields. In Ph-BTBT-C10 semiconductors, both types are important for understanding charge dynamics, conductivity and carrier transport mechanisms. Experimental confirmation compared to theoretical models is useful to understand high-field charge dynamics, which in turn is necessary for optimising the design and performance of semiconductor devices.

Keywords: Ph-BTBT-C10, TCAD modelling, device physics, density of state, Pool-Frenkel mobility

1. Introduction

The area of organic electronics is growing with organic thin film transistors (OTFTs). In this case, the organic electronics will require at least equally effective as inorganic semiconductors (OSC) for disposable electronics, smartcards, light-emitting diodes, signage and wearable technology, sensors, etc. [1, 2]. However, to make organic electronics a competitive option to in OSCs, in these limited application areas the manufacturing process should be simple as well as inexpensive. Ideally, techniques such as solution-based deposition of OSCs [3] onto plastic substrates would be

IntechOpen

utilised, with manufacturing taking place in ambient conditions and at room temperature. Although in comparison with flexible forms of inorganic semiconductors like ultra-thin single-crystal silicon [4], OTFTs must exhibit good electro-mechanical stability so that they remain flexible enough but still retain their electrical performance. On the other hand, achieving cost-effective production on flexible plastic sheets along with mechanical durability could open doors to innovative applications harnessing the unique attributes of carbon-based materials.

A better comprehension of the OTFT device characteristics is needed to enable improvement in the development of efficient solution-processable organic semiconductors for wider applications. In this process, device models are vital because they connect macroscopic observables such as current densities to the microscopic molecular level properties of OSCs. However, creating accurate and reliable models for OTFTs presents significant challenges. Disorder, mobility anisotropy, various factors including traps or grain boundaries and film morphology details also affect charge transport processes. Integrating all these aspects into a single OTFT model is complex. Moreover, an accurate OTFT model must consider the specific microscopic charge-transport mechanisms at play. Many times, in particular for solution-processed ones, charge transport occurs in OTFTs mainly *via* a hopping mechanism where electronic couplings between adjacent OSC molecules or polymer chain segments are relatively weak. Understanding and modelling the other effect of charge movement is important to progress solution-processable OTFTs towards being used in flexible electronics applications with optimal performance levels. Current unified models for OTFTs often struggle to accommodate the newly developed OSCs. This makes device modelling crucial for understanding the density of states (DOS) [5] and charge transport in these materials. To meet the rising demand for OSC-based devices, we present a study on the modelling of Ph-BTBT-C10. Also, the existing unified models for OTFTs fail to accommodate newly synthesised OSCs. Therefore, device modelling is essential to understand the density of states (DOS) and charge transport [6, 7].

2. Physics-based technology computer-aided design (TCAD) modelling

A physics-based TCAD (Technology Computer-Aided Design) model consists of a detailed representation of semiconductor devices that captures the fundamental physical processes governing their behaviour. This involves modelling mobility and recombination-generation processes as well as considering material features like doping level, defect and band structure. Another area of interest in the model covers electrostatics, which involves solving Poisson's equation to analyse electric fields and potential distribution across the device. Furthermore, it has built-in thermal effects, quantum phenomena, and other physical interactions necessary for an exact device performance under different conditions. SILVACO ATLAS [8] is used to simulate the model, and it solves semiconductor material parameters by applying Maxwell's equations to 2D/3D mesh points and solving the Poisson equation for boundary conditions [9]. It is possible to develop both AC and DC transient models for any device using this methodology.

2.1 Semiconductor fundamental modelling

The functioning of a transistor, which switches back and forth between on and off states, is mainly influenced by the properties of the semiconductor materials employed. As an illustration, the compound 2-decyl-7-phenyl-benzothienobenzothiophene

(Ph-BTBT-10) [10] possesses both sigma (σ) and pi (π) bonds, as represented in **Figure 1(a)**. The weak π bond gets formed by the electrons located farther from the nucleus, p orbitals. The alternating sequence between pi and sigma bonding results in shorter pi bonds than their conventional sizes. This change leads to partial localisation of π-bonds over specific atoms that then initiate conductivity in the OSC layer [see **Figure 1(b)**]. This band structure gap created by alternating bonding/anti-bonding interactions depends on a particular arrangement of organic molecules.

A molecule without charge is the cornerstone of conjugated polymers, which, depending on their specific features, could be insulators, semiconductors or conductors. These polymers are made up of a linear chain of carbon atoms. For a chain of 'N' atoms spaced by distance 'd', when N is large, the total length of the chain is approximately 'Nd'. According to the quantum-mechanical model for a free particle in a one-dimensional box with zero potential inside and infinite outside, the energy levels are given by:

$$E_n = \left(\frac{\hbar^2}{8m}\right) \times \left(\frac{n^2}{(Nd)^2}\right), \text{with } n = 1, 2, 3 \dots \quad (1)$$

where $\hbar$ is the reduced Planck's constant, m is the electron mass, and n is the quantum number (n = 1, 2, 3 ...). The three-dimensional spacing 'd' can be computed as a 3D matrix. It has been assumed that the 'N'-p orbitals contribute two electrons per molecule to form pi bond. The Highest Occupied Molecular Orbital (HOMO) energy was determined using eq. (2); while eq. (3) shows the lowest unoccupied molecular orbital (LUMO) energy:

$$E(HOMO) = \left(\frac{\hbar^2}{8m}\right) \times \left(\frac{(N/2)^2}{(Nd)^2}\right) \quad (2)$$

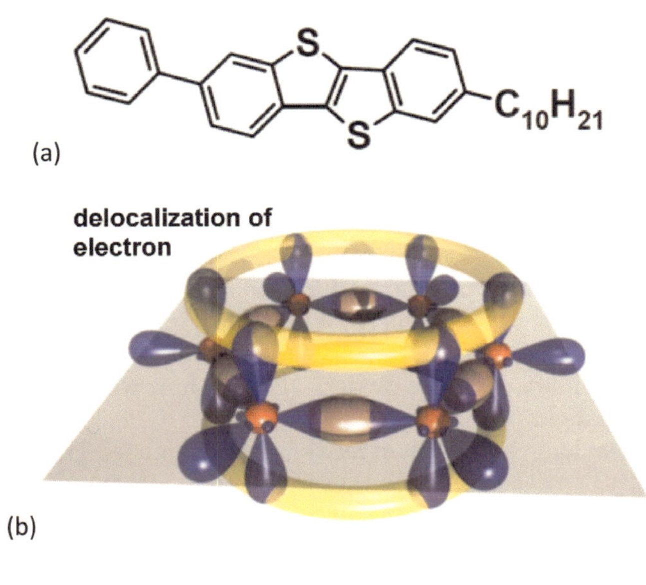

Figure 1.
(a) Chemical structure of Ph-BTBT-C10 (b) Delocalisation of π electron.

$$E(LUMO) = \left(\frac{\hbar^2}{8m}\right) \times \left(\frac{(N/2+1)^2}{(Nd)^2}\right) \tag{3}$$

Thus, band gap of OSC could be represent as eq. (4)

$$E_G = E(LUMO) - E(HOMO)$$

$$E_G = \left(\frac{h^2}{8m}\right) \times \left(\frac{N+1}{(Nd)^2}\right) \approx \left[h^2/8md^2\right]/N \text{ for large } N \tag{4}$$

Devices must be built and optimised with a good grasp of the conductance characteristics. Holes and electrons move due to both electric fields and concentration gradients. Hole concentrations are key charge carrier in p-type semiconductors. The behaviour is defined by the equation for hole density current.

$$\vec{J}_p = qp\mu_p\vec{E}_p - qD_p\nabla p \tag{5}$$

It is providing an inclusive approach for current density (J) in a conductive material. Drift on the other hand is determined by two factors; drift current density and diffusion current density. 'The drift current' density, denoted as $qn\mu E$ related charge carriers, which moves under the field E represented as J n = qnμ E, where the mobility of charge carriers is μ, the charge of carriers is q and n is the carrier density. In addition, diffusion constant D depends on the features of a material. Diffusion current density (qD∇n) shows charge carriers movement caused by concentration gradients. There is a difference in charge-carrier density within the material, represented by the gradient ∇n, charge carriers naturally move from areas of higher concentration to lower and create a diffusion current. The diffusion constant D, which is a key factor in this process, depends on the specific properties of the material and can be defined as follows:

$$D_n = \frac{kT}{q}\mu_n; D_p = \frac{kT}{q}\mu_p \tag{6}$$

The presence of an electric field influences the drift velocity (μ) of charge carriers in a conductor. However, since it depends on the properties that material is made of, or rather its diffusion constant (D), the previous model simulation output did not give us an I_{DS} vs V_{GS} graph as expected. This discrepancy between experimental and simulated results implies that further work is required to understand the diffusion constant (D) in more detail.

In this formula [11]; D represents the diffusion constant, while k stands for Bolzmann's constant that connects energy with temperature. The letter T stands for absolute temperature measured in Kelvin, q is the charge carrier such as electron or hole, and μ is the mobility of charge carriers, showing how well they can move under influence of electric field.

The formula that known as Einstein relation for drift model. The current (I) can be related to the diffusion coefficient (D) thus leading into Einstein's relationship concerning drift model in a conductor. For instance, we can express the current (I) through a conducting medium:

$$I = n * q * D * A \tag{7}$$

where n is the number of charge carriers per unit volume. A represents the cross-sectional area of conductor. When a conductor is placed to an electric field (E), the charge carriers experience a force (F) given by

$$F = q * E \tag{8}$$

Considering the force equation and the definition of mobility (μ), we get:

$$D = \mu * E \tag{9}$$

Substituting this expression for D into the current equation, we obtain

$$I = n * q * \mu * E * A \tag{10}$$

Now, KT in the original formula, probably represents the thermal energy available to the charge carriers due to the temperature (T). By considering this thermal energy, we can modify the formula to:

$$D = (KT/q) * \mu \tag{11}$$

The drift velocity of charge carriers is determined by this equation in relation to the electric field that is applied [12, 13]. For instance, Fermi-Dirac distribution function and drift velocity are important factors in charge-carrier transport in Ph-BTBT-C10 semiconductors. Once an electric field is applied across a material, it gives rise to a force on the charges (charge carriers); thus, they move with specific velocities. To establish how charge carriers are distributed over various energy levels, especially those near the Fermi level, one uses the Fermi-Dirac distribution function. The standard drift coefficient D cannot capture correctly the current density for higher values of carrier concentration though. In this case, however, the Fermi-Dirac integral order ½ should be expressed using some other form for calculating fermions, as will be seen from the integration of all energy states (E) with respect to a given fermionic distribution called Fermi-Dirac integral:

$$D_p = \frac{\left(\frac{KT_L}{q}\mu_p\right)F_{\frac{1}{2}}\left\{\frac{1}{KT_L}\left[E_{Fp} - E_v\right]\right\}}{F_{-\frac{1}{2}}\left\{\frac{1}{KT_L}\left[E_{Fp} - E_v\right]\right\}} \tag{12}$$

$$E_{Fp} = -q1\,\phi\,(x) \tag{13}$$

The charge-potential relationship is represented by the term (E_{Fp}), which is the electric potential energy (E) of a charged particle inside an electric field. Here are the parts of this equation:

- E_{Fp}: This represents the electric potential energy in joules (J) and shows how much energy a charged particle has because it is located in an electrostatic field.

- q: The charge, as measured in coulombs (C), is indicative of the amount of electrical charge carried by the particle. A positive (q) indicates a positive charge and negative (q) implies a negative charge.

- ϕ (x)**:** The electric potential or voltage in volts (V), signifies the electrical potential at some place inside the electric field. It denotes how much potential energy per unit of charge exists at that point.

If a charged particle is moved against an electric field from a lower potential region to a higher potential one, the work done is generally given a negative sign. This convention reflects the negative change in potential energy, indicating that work is being done on the particle against the direction of the field it is moving in. At low temperatures, eq. (8) describes fermions – particles with half-integer spins. It is through this equation that scientists understand the quantum mechanical aspects of materials. In agreement with Pauli exclusion principle at zero Kelvin absolute zero, Fermions are found to occupy the lowest available energy states in any system and also no two identical Fermions can occupy the same quantum state. The Fermi-Dirac distribution function tells how often fermion fills a particular energy state [14–17].

The Fermi-Dirac integral of order 1/2 comes about when integrating this distribution function over all possible energies. In contrast to Fermi-Dirac distribution, which takes discrete values and looks like a step-wise function, Fermi-Dirac integral of order 1/2 is continuous and defined as follows:

$$f_{1/2} = \int_{E_V}^{\infty} \frac{E^{\frac{1}{2}}}{(1 + exp\,(E - E_F))}\,dE \tag{14}$$

The model's accuracy improved significantly when charge density was incorporated as a function of the Fermi-Dirac integral of order ½. By adding this function along with the drift coefficient, the model better aligned with the experimental data. This approach ensures that the theoretical predictions fall within the experimental range. The next section details the parameter adjustments made to achieve this enhanced accuracy.

In this TCAD model eq.(15) is defined for OSC band gap [13], Here E_{g0} is a theoretical band gap [10] and it could be extracted also by experimental work at room temperature. $'T'$ is the temperature variable [18] and E_{G_α} and E_{G_β} are fitting variables [19]:

$$E_g(T_L) = E_{g0} + E_{G_\alpha}\left[\frac{300^2}{300 + E_{G_\beta}} - \frac{T^2}{T + E_{G_\beta}}\right] + \Delta E_g \tag{15}$$

$$\Delta E_g = -\zeta_C\left[\frac{\varepsilon_s^5}{N}\left(m_0 + \zeta_B * T^2\frac{\varepsilon_s}{N}\right)\right]^{-\frac{1}{4}} \tag{16}$$

Here, m_0 represents the mass of electrons at static conditions (9.11×10^{-31}kg). The parameters ζ_B and ζ_c are fitting parameters. N denotes the doping concentration, and ε_s stands for the semiconductor dielectric constant [20]. The effect of doping on the band gap and charge transport in OTFTs is generally more significant in p-type OSCs. By introducing electron-accepting dopants or oxidation of OSC, such small molecules or other organic materials, into the OSC layer, the band gap can be altered. The presence of dopants creates charge carriers, either holes or polarons, which can significantly affect the charge transport and mobility in the OSC [20].

2.2 DOS of mobile charge carrier

These equations can be integrated over the Fermi-Dirac [21] statistics that are parabolic in nature and occur in both valence and conduction bands whose energies are E_C (at the bottom of the conduction band) and E_V (at the top of the valence band). These give eqs. (17, 18).

The Fermi-Dirac integral, a mathematical function, quantifies the likelihood of an electron occupying a state with energy E. At the Fermi level, the occupation probability reaches 50%. The effective mass characterises the inertia of an electron or hole, accounting for its interactions with the semiconductor lattice, distinct from its actual mass. The Boltzmann constant serves as a fundamental link between energy and temperature.

The temperature dependence of electron and hole concentrations arises from the Fermi-Dirac integral [22]. With increasing temperature, the probability of occupation for states near the Fermi level rises, resulting in heightened electron and hole concentrations. By utilising the formulas for electron and hole concentrations, one can compute the carrier concentration within a semiconductor material. This parameter plays a vital role in determining the material's electrical properties:

$$\boldsymbol{n = N_C\ exp\left(\frac{E_F - E_{CB}}{kT_L}\right)} \tag{17}$$

$$p = N_V\ exp\left(\frac{E_{VB} - E_F}{kT_L}\right) \tag{18}$$

$$N_C(T_L) = 2\left(\frac{2\pi m_e^* kT_L}{h^2}\right)^{\frac{3}{2}} = \left(\frac{T_L}{300}\right)^{NC\cdot F} NC300 \tag{19}$$

$$N_V(T_L) = 2\left(\frac{2\pi m_h^* kT_L}{h^2}\right)^{\frac{3}{2}} = \left(\frac{T_L}{300}\right)^{NV\cdot F} NV300 \tag{20}$$

Here, n & p signifies electron ad hole concentration, N_C and N_v (eqs. 19, 20) is DOS in the conduction band and valence band [23], $f_F(E)$ is the probability of an electron in the conduction band and also the probability of a hole in the valence band. The probability of an electron in the valence band and a hole in the conduction band is $(1 - f_F(E))$, m∗ signifies the effective mass of the hole or electron, k signifies Boltzmann's constant, T_L is the temperature variable in the entire model, and E_F is the Fermi level. The values of parameters are collected from the literature survey, and some are adjusted from iteration of simulation.

2.3 DOS for acceptor and donor

In theory, both positive and negative charges can be mobile in a semiconductor film. However, in practice, most semiconductors usually have only one kind of charge that is movable. As a result, contacts often inject only one type of charge, which may end up getting trapped. Due to the wide energy band gap, the thermal generation of charge carriers is rather feeble [24]. Consequently, it is often assumed that every molecule carries only a single free charge.

That total DOS is often equal to the concentration of the molecules $(10^{20} - 10^{21})$. The DOS has been described for two charge carriers, the first mobile charge carrier (hold & electron) and the second static charge carrier (acceptor and donor). The DOS is assumed to consist of four bands: two tail bands (representing a donor-like valence band and an acceptor-like conduction band) and two deep-level bands (one being acceptor-like and the other donor-like). These bands are modelled using Gaussian distributions:

$$g(E) = g_{TA}(E) + g_{TD}(E) + g_{GA}(E) + g_{GD}(E) \tag{21}$$

$$g_{TA}(E) = N_{TA}\ \exp\left[\frac{E - E_c}{W_{TA}}\right] \tag{22}$$

$$g_{TD}(E) = N_{TD}\exp\left[\frac{E_v - E}{W_{TD}}\right] \tag{23}$$

$$g_{GA}(E) = N_{GA}\ \exp\left[-\left[\frac{E_{GA} - E}{W_{GA}}\right]^2\right] \tag{24}$$

$$g_{GD}(E) = N_{GD}\exp\left[-\left[\frac{E - E_{GD}}{W_{GD}}\right]^2\right] \tag{25}$$

In terms of the DOS description [25, 26], there are two different distributions considered: an exponential tail distribution and a Gaussian distribution. Each distribution has specific parameters that define the DOS. The DOS can be described using two different distributions. For an exponential tail distribution, the DOS is characterised by the intercept densities at the conduction and valence band edges (N_{TA} and (N_{TD}) and the characteristic decay energies (W_{TA} and W_{TD}) as in eq. (22, 23). For Gaussian distributions, the DOS is characterised by the total DOS (N_{GA} and N_{GD}), the characteristic decay energies (W_{GA} and W_{GD}), and the energy of peak distributions (E_{GA} and E_{GD}) as in eqs. (24, 25).

The ionised densities of acceptor and donor-like states (p_T and n_T, respectively) are given by eqs. (26, 27):

$$p_T = p_{TA} + p_{GA} \tag{26}$$

$$n_T = n_{TD} + n_{GD} \tag{27}$$

where p_{TA}, p_{GA}, n_{TD} and $n_{GD}$are given below

$$p_{TA} = \int_{E_V}^{E_C} g_{TA}(E) \cdot f_{t_{TA}}(E, n, p)dE \tag{28}$$

$$p_{GA} = \int_{E_V}^{E_C} g_{GA}(E) \cdot f_{t_{GA}}(E, n, p)dE \tag{29}$$

$$n_{TD} = \int_{E_V}^{E_C} g_{TD}(E) \cdot f_{t_{TD}}(E, n, p)dE \tag{30}$$

$$n_{GD} = \int_{E_V}^{E_C} g_{GD}(E) \cdot f_{t_{GD}}(E, n, p)dE \tag{31}$$

Ionised density is a very significant aspect of semiconductor physics. What this means is the number of ionised impurity atoms present in the material. Semiconductors are intentionally doped with impurities to alter their electrical properties. They are divided into two categories, donors and acceptors, according to how they affect the conductivity of the semiconductor. When a dopant atom enters the lattice, it can donate or accept an electron from/to either the valence band or conduction band, respectively. This process results in the creation of charged particles called ionised dopants. The p_T refers to the density of positive ions while n_T stands for negatively charged ones. In most cases, these quantities are inversely related; if p_T is high then n_T will be low and vice versa. Essentially, this occurs because semiconductors generally have very few total free charge carriers so the presence of one species of ionised dopant tends to suppress the other type's availability.

Temperature and an electric field can modify the dopants ionised density. Normally, this density rises with the increase in temperature and decreases in the direction of electric field. The ionisation of dopants and the consequent ionised density have a vital role to play in determining the electrical response of semiconductor devices [27]. By controlling concentration and distribution of dopants, semiconductor engineers are able to shape material conductivity and other electrical properties into materials that fit specific applications best.

In eq. (28–31), $f_{t_{TA}}(E, n, p)$ and $f_{t_{GA}}(E, n, p)$ are the ionisation probabilities for the tail and Gaussian acceptor DOS, while $f_{t_{TD}}(E, n, p)$ and $f_{t_{GD}}(E, n, p)$ are the ionisation probabilities for the donors and are described as eq. (32–35):

$$f_{t_{TA}}(E, n, p) = \frac{v_n \sigma_{TAE} n + v_p \sigma_{TAH} n_i \exp\left[\frac{E_i - E}{kT}\right]}{v_n \sigma_{TAE}\left(n + n_i \exp\left[\frac{E - E_i}{kT}\right]\right) + v_p \sigma_{TAH}\left(p + n_i \exp\left[\frac{E_i - E'}{kT}\right]\right)} \tag{32}$$

$$f_{t_{GA}}(E, n, p) = \frac{v_n\ \sigma_{GAE}\ n + v_p\ \sigma_{GAH}\ n_i \exp\left[\frac{E_i - E}{kT}\right]}{v_n \sigma_{GAE}\left(n + n_i \exp\left[\frac{E - E_i}{kT}\right]\right) + v_p \sigma_{GAH}\left(p + n_i \exp\left[\frac{E_i - E_0}{kT}\right]\right)} \tag{33}$$

$$f_{t_{TD}}(E, n, p) = \frac{v_p \sigma_{TDH}\ p + v_n \sigma_{TAE}\ n_i \exp\left[\frac{E - E_\varsigma}{kT}\right]}{v_n \sigma_{TAE}\left(n + n_i \exp\left[\frac{E - E_i}{kT}\right]\right) + v_p \sigma_{TDH}\left(p + n_i \exp\left[\frac{E_i - E'}{kT}\right]\right)} \tag{34}$$

$$f_{t_{GD}}(E, n, p) = \frac{v_p \sigma_{GDH}\ p + v_n \sigma_{GDE}\ n_i \exp\left[\frac{E - E_i}{kT}\right]}{v_n \sigma_{GDE}\left(n + n_i \exp\left[\frac{E - E_i}{kT}\right]\right) + v_p \sigma_{GDH}\left(p + n_i \exp\left[\frac{E_i - E^-}{kT}\right]\right)} \tag{35}$$

$$v_t\ (\text{Thermal Velocity}) = \sqrt{\frac{3kT}{m}} \tag{36}$$

n semiconductor physics, the thermal velocity of the electron [28], is often called 'vn'. This defines the average speed at which an electron moves due to thermal energy within it. When the temperature increases, the kinetic energy of the electrons also increases, due to which their heat velocity also increases. It is one of the important parameters for understanding the dynamics of electronic transport like diffusion and drift, both of which are critical to device operations in semiconductors. Similarly, vp exhibits the thermal motion of holes in a semiconductor. Holes can be considered as vacancies in valence band filled by surplus electrons that act as if they were positively charged particles.

Like electrons, thermal energy is also present in holes due to the nearby temperature, which leads to thermal movement by which hole transport mechanisms can be analysed and general charge transfer characteristics in semiconductors explained. 'Ni' which is the intrinsic carrier concentration is an important characteristic of any semiconductor material. This indicates the number at equilibrium of charge carriers (electrons and holes) in absence of any external dopants. In equilibrium state, generation and recombination rates balance each other [29], leading to constant carrier concentration. Intrinsic carrier concentration relies on the energy bandgap as well as the temperature of a material and can be used as a reference point for how doping affects carrier concentrations.

The electron capture cross-sections [30], σ_{TAE} and σ_{GAE} are parameters that indicate how likely electrons can be captured by acceptor-like states in semiconductors. These electron capture cross-sections are used to determine the trapping of electrons in these acceptor states and hence influence carrier dynamics as well as recombination processes across the material. Acceptor tail and Gaussian states stand for distinguishable energy configurations of acceptor state while their respective capturing cross-sections give a hint on the behaviour of this phenomenon.

Additionally, hole-capture cross-sections could also be represented as σ_{TAH} and σ_{GAH} stand for probabilities holes can get trapped by acceptor-like states that are found within semiconductors. Moreover, these trap cross-sections play a vital role in helping one to comprehend hole trappings coupled with recombination mechanisms The acceptor tail and Gaussian states offer alternative energy landscapes for holes to associate with before they reach an agreement with the respective capture probabilities. As far as donor states are concerned; the representative symbols include σ_{TDH}, σ_{GDE}, and δ_{GDH}. Inserting impurity atoms into the crystal lattice so as to bring about excessive valence electrons results in donor levels. Donor state's capture cross-sections describe the chances of both electrons and holes being trapped by these donor-like states. These are also important in charge-carrier-donor state interactions, which affect carrier lifetime, recombination rates and other semiconductor characteristics.

Semiconductor device design and analysis require an understanding of ionisation probabilities, thermal velocities, intrinsic carrier concentration and capture cross-sections. Carrier transport, recombination processes and overall device performance are directly influenced by these parameters. The studies can also help in optimising the properties of devices for better performance and increased efficiency besides maximising electronic technology development. For a steady-state situation in a semiconductor material, there is a balance between the net rates of generation/recombination of electrons (R_n) and holes (R_p) (eq. 37). This means that electrons recombine at the same rate as holes are generated so that the concentration of carriers remains constant in any material under consideration. The latter form is important for maintaining stable electrical properties within semiconductor devices:

$$R_{n,p} = \int_{E_V}^{E_C} \left(\frac{v_p v_n \sigma_{TAH} \sigma_{TAE} (np - n_i^2) g_{TA}(E)}{\sigma_{TAE} v_n \left(n_i \exp\left[\frac{E-E_i}{kT}\right] + n\right) + \sigma_{TAH} v_p \left(n_i \exp\left[\frac{E_i-E}{kT}\right] + p\right)} + \frac{v_p v_n \, \sigma_{GAH} \sigma_{TAE} (np - n_i^2) g_{GA}(E)}{v_n \sigma_{GAE} \left(n_i \exp\left[\frac{E-E_i}{kT}\right] + n\right) + v_p \, \sigma_{GAH} \left(n_i \exp\left[\frac{E_i-E}{kT}\right] + p\right)} + \frac{v_p v_n \, \sigma_{TDH} \sigma_{TDE} (np - n_i^2) g_{TD}(E)}{v_n \sigma_{TDE} \left(n_i \exp\left[\frac{E-E_i}{kT}\right] + n\right) + v_p \sigma_{TDH} \left(n_i \exp\left[\frac{E_i-E}{kT}\right] + p\right)} + \frac{v_p v_n \sigma_{GDE} \sigma_{GDH} (np - n_i^2) g_{GD}(E)}{v_n \sigma_{GDE} \left(n_i \exp\left[\frac{E-E_i}{kT}\right] + n\right) + v_p \sigma_{GDH} \left(n_i \exp\left[\frac{E_i-E}{kT}\right] + p\right)} \right) dE \quad (37)$$

The values of f_t are determined by eqs. (32) through (36). The Shockley-Read-Hall (SRH) recombination/generation rate due to defect states is derived in Ref. [31]. The rate at which the carriers (electrons or holes) interact with defect states in the semiconductor material is referred to as SRH recombination/generation rate. Defect states can come from impurities, crystal lattice imperfections and other structural irregularities. **Table 1** shows the values for acceptor & donor DOS parameter for Ph-BTBT-C10.

Parameters	Values	Unit	Parameters	Values	Unit
N_{TA}	9.0×10^{12}	cm^{-3}/eV	σ_{TAE}	2.2×10^{-16}	cm^2
N_{GA}	2.2×10^{12}	cm^{-3}/eV	σ_{TDE}	1.7×10^{-14}	cm^2
N_{TD}	4.5×10^{12}	cm^{-3}/eV	σ_{GAE}	3.0×10^{-16}	cm^2
N_{GD}	2.2×10^{12}	cm^{-3}/eV	σ_{GDE}	1.2×10^{-14}	cm^2
E_{GA}	0.5	eV	σ_{TAH}	2.5×10^{-14}	cm^2
E_{GD}	0.7	eV	σ_{TDH}	1.5×10^{-16}	cm^2
W_{TA}	0.30	eV	σ_{GAH}	3.5×10^{-14}	cm^2
W_{TD}	0.50	eV	σ_{GDH}	1.2×10^{-16}	cm^2
W_{GA}	0.15	eV			
W_{GD}	0.15	eV			

Table 1.
Acceptor & donor DOS parameter values for Ph-BTBT-C10.

2.4 Mobility modelling

In disordered semiconductors, the charge transportation accrues by more than one phenomenon [32]. Unlike crystalline semiconductors, where charge carriers pass along well-defined energy bands, in disordered semiconductors, the energy bands are disturbed, resulting in localised energy states. These localised states, sometimes referred to as trap sites or jumping sites, are scattered throughout the material. From these taps charges flow source to drain, it is described as hopping charge transportation.

Hopping transport happens through two basic mechanisms: thermal activation and quantum tunnelling. In thermal activation, charge carriers get sufficient energy from the electric field or heat to overcome the energy barrier between hopping sites, as shown in **Figure 2**. Quantum tunnelling [33], on the other hand, includes charge carriers flowing through energy barriers without obtaining the requisite energy due to their wave-like nature.

The rate of hopping charge transfer in TFTs relies on parameters [34] such as temperature, the density of trap sites, and the degree of disorder in the semiconductor material. Higher temperatures and larger trap site densities often result in more substantial hopping conduction, influencing the TFT's electrical properties.
While hopping charge transport has certain disadvantages, it can restrict carrier mobility and lower device performance compared to crystalline semiconductors. Researchers are continually working on enhancing the characteristics of disordered semiconductors and investigating novel materials to boost their performance:

$$\mu_{n_{hop}} = \frac{qV_{0nhop}}{kT_L}\left[\int_{-\infty}^{E_{trn}} g_a(E)dE\right]^{-\frac{2}{3}}$$
$$\exp\left[-2\left(\frac{3\beta_{n\_hop}}{4\pi}\right)^{1/3}\gamma_{n\_hop}\left[\int_{-\infty}^{E_{trn}} g_a(E)dE\right]^{-1/3}\right] \quad (38)$$
$$\mu_{p_{hop}} = \frac{qV_{0phop}}{kT_L}\left[\int_{-\infty}^{E_{trn}} g_d(E)dE\right]^{-\frac{2}{3}}$$

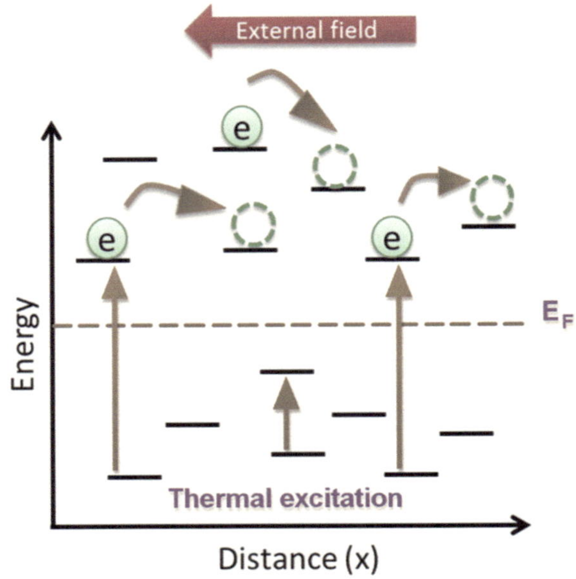

Figure 2.
Electron migration from lower to higher states and under hopping transport.

$$\exp\left[-2\left(\frac{3\beta_{p\_hop}}{4\pi}\right)^{1/3}\gamma_{p\_hop}\left[\int_{-\infty}^{E_{trn}} g_d(E)dE\right]^{-1/3}\right] \quad (39)$$

The hopping model is described in ref. [35]. In this model, the effective mobility of electrons and holes is indicated as $\mu_{n_{hop}}$ and $\mu_{p_{hop}}$ accordingly. Eq. (38) refers to electrons, where V_{0nhop} reflects the frequency of electrons trying to jump within the conduction band, and $g_a(E)$ symbolises the DOS for the acceptor-like conduction band. The variables $\beta_{n_{hop}}$ and $\gamma_{n_{hop}}$ indicate the percolation constant and the reciprocal of the carrier localisation radius, respectively. The carrier is likely to transition from its current occupied state to a hopping site situated within the effective transport level of energy E_{tr}.

Similarly, in eq. (39), the subscript 'p' signifies holes, and the parameters coincide with the prior description for holes in the hopping mobility model. These equations represent the behaviour of charge carriers in a disordered or amorphous material, where charge transport occurs by hopping between localised states [36, 37].

Initial simulations showed the differences between the hopping mobility model and collected experimental data [38]. To better represent the high mobility of charge carriers and understand more about the electrical behaviour of materials, a model that can address higher mobility cases must be adopted. The search for more appropriate frameworks is prompted by the need to understand the characteristic behaviours of charge carriers at high mobility levels correctly. This entails thinking about a model that covers all factors other than thermal activation such as Poole-Frankel mobility models. We would like to get a clear picture of carrier motion and gain a deeper understanding of the electric behaviour of substances so as to obtain a true depiction of charge carrier velocity [37].

Poole-Frenkel mobility model, which was proposed by Charles A Poole and Frenkel in 1938, gives a useful framework to understand changes in the mobility of charge carriers such as electrons or holes within dielectric materials when electric field is applied to them. The Poole-Frenkel mobility model relies on the assumption that charge carriers are trapped in localised states due to inherent crystal defects in dielectric materials. These limited states hinder their free movement when subjected to an electric field. This force dislodges the charge carriers from their position of detention, enabling them to move through the material.

Further explanation of Poole-Frenkel mobility model will entail a thorough evaluation of its major constituents. Defects in the material's crystal structure that are localised indicate the existence of charge carriers trapping sites. The difference in energy levels between the trapped state and the conduction band is represented by a barrier. Before moving freely through matter, charge carriers have to overcome it. The presence of an electric field reduces potential energy barriers mostly because of how fields interact. As a result, there is less resistance for charge carriers as they go across lower barriers. Consequently, with stronger electric fields pulling the charge carriers out of their localised states into conduction bands, their mobility is boosted. A complete model of Poole-Frenkel conduction is defined in ref. [39]. The mobility generated by these effects is formulated by the below functions:

$$\mu_{n\_pf}(E) = \mu_n \exp\left(\left(\frac{\beta_{n_{pf}}}{kT_{neff}} - \gamma_{n_{pf}}\right)\sqrt{|E|} - \frac{\delta_{n_{pf}}}{kT_{neff}}\right) \tag{40}$$

$$\mu_{p\_pf}(E) = \mu_p \exp\left(\left(\frac{\beta_{p_{pf}}}{kT_{peff}} - \gamma_{p_{pf}}\right)\sqrt{|E|} - \frac{\delta_{p_{pf}}}{kT_{peff}}\right) \tag{41}$$

$$T_{peff} = \frac{TOP \cdot PFMOB.T_L}{TOP \cdot PFMOB - T_L} \tag{42}$$

The Poole-Frenkel mobility model centres upon the dynamics of charge motion, as elucidated through eq. (40) to (42). These equations encapsulate the effective mobility of positive and negative charges, denoted as μ_{npf} and $\mu_{p_{pf}}$ respectively. Within these equations, μ_{n0} signifies electron mobility, while μ_{p0} represents hole mobility. The terms $\delta_{p_{pf}}$ and $\delta_{p_{pf}}$ are indicative of the activation energy pertinent to holes and electrons, respectively, and are evaluated under zero electric field conditions. Further insight is provided by $\beta_{p_{pf}}$ and $\beta_{n_{pf}}$, denoting the Poole-Frenkel factor for holes and electrons. Within this context, the model encompasses net effective temperatures, specifically $T_{p_{eff}}$ and $T_{n_{eff}}$, which pertain to holes and electrons correspondingly. These variables collectively contribute to a comprehensive depiction of charge-carrier mobility and behaviour within the context of the Poole-Frenkel mobility model [16].

3. Simulation and verification of model

In ref. [10], Hiroaki Iino et al. fabricated a bottom-gate, top-contact OTFT carried out using polycrystalline thin films of Ph-BTBT-10 on SiO2/Si substrates. The gate oxide was a standard 300 nm-thick SiO2 layer. To achieve the active phase of the semiconductor. The active semiconductor layer of Ph-BTBT-10, about 50 nm thick, was deposited on these substrates. Channel dimensions were defined to be 100 μm in length and 500 μm in width. These devices aimed to investigate whether Ph-BTBT-10 polymer could be used further as an active material in solution-processable OTFTs. The output ($I_{DS} - V_{DS}$) and transfer ($I_{DS} - V_{GS}$) measurements characterisation of the fabricated devices are provided. The experiment design achieved the objective of investigating Ph-BTBT-10 as a possible means to achieve defect-free and molecularly smooth films. This resulted in samples that present predictable performance within different devices with only a few changes. Also, the Ph-BTBT-10 films showed good thermal stability including operation temperatures of up to 200°C.

Based on this description, the device structure, as shown in **Figure 3** [40], is defined in ATLAS SILVACO. The simulation focused on evaluating the band gap and DOS while storing temporary log and str files. The amount of charge flow is determined using the charge transport model.

Silvaco ATLAS simulation generates a graph between drain current (I_{DS}) and gate voltage (V_{GS}) along with drain voltage (V_{DS}). The results are being compared to experimental data to confirm the correctness of the simulation. **Figure 4** presents the I_{DS} vs. V_{GS} graph in both logarithmic and ordinary scales. The logarithmic scale offers a clearer picture of the data at low current values, when large changes may occur. The graph presently displays how the drain current fluctuates with different gate voltage values. Meanwhile, the normal scale gives a larger perspective of the I_{DS} vs. V_{GS} relationship throughout the complete range of gate voltage values [10]. **Figure 5**

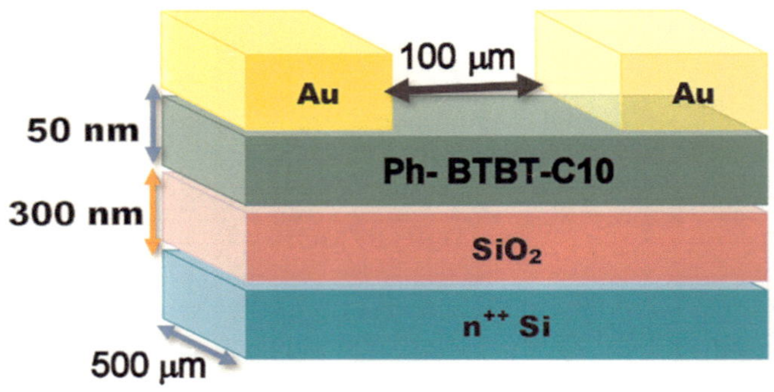

Figure 3.
Device structure of Ph-BTBT-C10 OTFT.

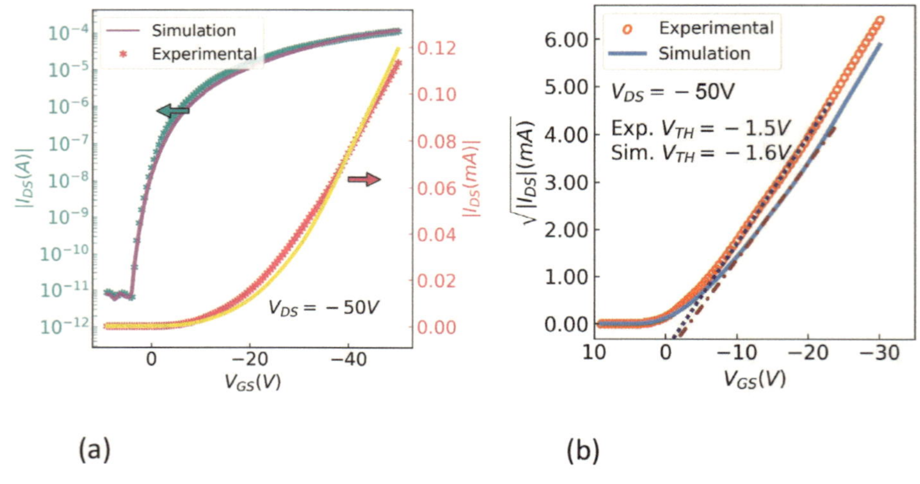

Figure 4.
Simulated and experimented with data of Ph-BTBT-C10 OTFT. (a) I_{DS} vs V_{GS} *characteristics in linear and log scale (b) Threshold voltage extraction.*

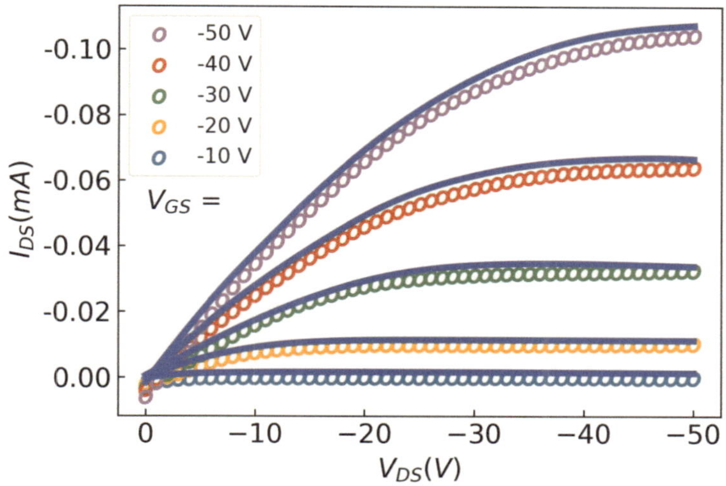

Figure 5.
Comparison of simulated and experimental data of I_{DS}vs V_{DS} *for Ph-BTBT-C10 OSC.*

Parameters	Experimental values	Simulated values	Unit
Capacitance per unit area (C_{ox})	10.4	10.8	nF/cm^2
Threshold Voltage (V_t)	−1.5	−1.6	volt
Transconductance (G_m)	$1.92x10^{-8}$	1.91×10^{-8}	siemens
Mobility in linear region (μ_{lin}) (at $V_{GS} = 0V$)	3.84×10^{-2}	3.83×10^{-2}	$cm^2/V - s$
Mobility in saturation region (μ_{sat}) (at $V_{GS} = -40V$)	1.12×10^1	1.02×10^1	$cm^2/V - s$
Sub-threshold Slope (SS)	7.81×10^{-1}	6.09×10^{-1}	Volt/ decade
On–Off ratio	10^7	10^7	(ratio)

Table 2.
Comparison of performance parameters of Ph-BTBT-C10 OTFT.

shows the comparison between the simulated and experimental data for I_{DS} versus V_{DS}. Comparison of performance parameters of simulated and experimental data is shown in **Table 2**. The objective is to guarantee that the simulated results closely match the experimental data to prove the dependability of the simulation model.

4. Channel length effect on device on–off ratio

Figure 6 shows forceful degradation of on–off ratio in 4 h-21 DNTT below the 60 μm. Because the channel length reduces, the power of the gate conductor to exclusively manage the charge distribution within the channel reduces and therefore the field applied from the S/D will influence the channel regime. Therefore, Vth roll-off also occurs in OTFT. In simulation, the OTFT is shown the SS (Sub-threshold) of

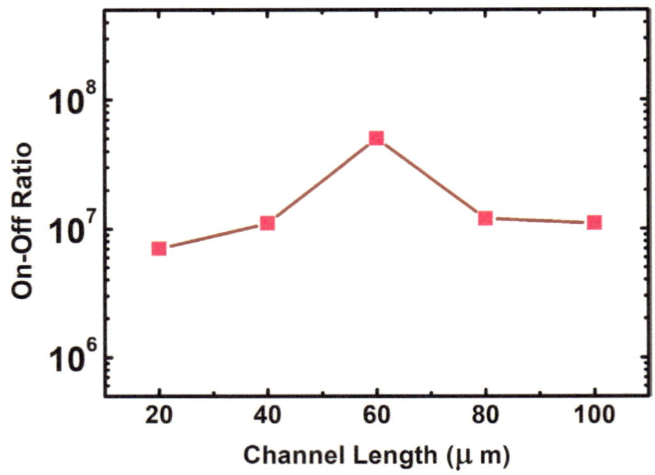

Figure 6.
Effect of channel length on On–Off ratio.

0.78 V/decade on 100 μm whereas 60 μm channel length exhibits relatively lower SS respectively. As a result, Fig. 8 shows forceful degradation of the on–off ratio in Ph-BTBT-C10 below 50 μm. Because the channel length reduces, the power of the gate conductor to exclusively manage the charge distribution within the channel reduces and therefore the field applied from the S/D will influence the channel regime. Therefore, Vth roll-off also occurs in OTFT. So, by the simulation, we get that 60 μm as the optimum value of channel length.

5. Conclusion

Conventional device physics has accomplished the TCAD modelling of the organic semiconductor ph-BTBT-C10. Accurate representation of the DOS was achieved using Gaussian distribution and tail distribution, and various transport mechanisms such as hopping, Poole-Frenkel and band-like conduction were included. The expression for the diffusion constant explained by Einstein's relation was updated with Fermi-Dirac-Distribution. As well as high-doping corrections were included in order to improve the model. The parameters were obtained in a couple of iterations and completed with values from either publications or other studies. This entire model was simulated on TCAD SILVACO and the results obtained were quite comparable in error to the actual experimental results acquired. The effect of the W/L ratio of the device was also studied and it was found that with a channel length of 60 microns, the structure works best for this configuration. These last findings testify as well to the reliability of the model and it is presented for further use for organic thin film transistors research.

Acknowledgements

The researchers are gratified to the DST Government of India for the financial support under the Early Career Research Award (ECRA) for Project No. ECR/2017/000179. Mr Shubham Dadhich is obliged for getting a Junior Research Fellowship (JRF) through this programme.

Author details

Shubham Dadhich[1*], Arun Dev Dhar Dwivedi[2] and Ram Babu Pareek[3]

1 Department of Electrical and Electronics Engineering, Poornima University Jaipur, Vidhani, India

2 School of Electronics Engineering, VIT University Vellore, Vellore, India

3 DESM, RIE Ajmer, NCERT, GOI, India

*Address all correspondence to: saytoshubham@gmail.com

References

[1] Zschieschang U, Klauk H. Organic transistors on paper: A brief review. Journal of Materials Chemistry C. 2019; **7**(19):5522-5533

[2] Dimitrakopoulos CD, Mascaro DJ. Organic thin-film transistors: A review of recent advances. IBM Journal of Research and Development. 2001;**45**(1): 11-27

[3] Lim B, Sun H, Lee J, Noh YY. High performance solution processed organic field effect transistors with novel diketopyrrolopyrrole-containing small molecules. Scientific Reports. 2017;**7**(1): 164

[4] Wang S, Weil BD, Li Y, Wang KX, Garnett E, Fan S, et al. Large-area free-standing ultrathin single-crystal silicon as processable materials. Nano Letters. 2013;**13**(9):4393-4398

[5] Dev Dhar Dwivedi A, Kumar Jain S, Dhar Dwivedi R, Dadhich S. Numerical simulation and compact modeling of thin film transistors for future flexible electronics. In: Vargas-Bernal R, He P, Zhang S, editors. Hybrid Nanomaterials—Flexible Electronics Materials [Internet]. London, UK: IntechOpen; 2020. [cited 2023 Jul 9]. Available from: https://www.intechopen.com/books/hybrid-nanomaterials-flexible-electronics-materials/numerical-simulation-and-compact-modeling-of-thin-film-transistors-for-future-flexible-electronics

[6] Boubaker A, Hafsi B, Lmimouni K, Kalboussi A. A comparative TCAD simulations of a P-and N-type organic field effect transistors: Field-dependent mobility, bulk and interface traps models. Journal of Materials Science: Materials in Electronics. 2017;**28**(11): 7834-7843

[7] Torricelli F. Charge transport in organic transistors accounting for a wide distribution of carrier energies—Part I: Theory. IEEE Transactions on Electron Devices. 2012;**59**(5):1514-1519

[8] SILVACO. Utmost IV User's Manual* (Document No. 51 2022). 2022. Available from: https://silvaco.com/wp-content/uploads/product/pdf/UtmostIV_brief.pdf

[9] Dwivedi ADD, Kumari P. Tcad simulation and performance analysis of single and dual gate OTFTS. Surface Review and Letters. 2020;**27**(05): 1950145

[10] Iino H, Usui T, Hanna J, ichi. Liquid crystals for organic thin-film transistors. Nature Communications. 2015;**6**(1):6828

[11] Xiao-Hong S, Jiu-Xun S, Chun-Hua X. Dependence of general Einstein relation on density of state for organic semiconductors. Organic Electronics. 2016;**35**:65-73

[12] Dadhich S, Mathur G, Dwivedi ADD. Mobile charge carrier based modeling of 4H–21 DNTT and structure analysis of OTFT. In: 2021 IEEE 2nd International Conference on Technology, Engineering, Management for Societal Impact Using Marketing, Entrepreneurship and Talent (TEMSMET) [Internet]. Pune, India: IEEE; 2021. pp. 1-5. Available from: https://ieeexplore.ieee.org/document/9768714/

[13] Dadhich S, Dwivedi ADD, Mathur G. Numerical simulation and analytical modelling of C8-BTBT-C8 organic transistor and analysis of semiconductor thickness. In: Tripathi A, Soni A,

Shrivastava A, Swarnkar A, Sahariya J, editors. Intelligent Computing Techniques for Smart Energy Systems [Internet]. Vol. 862. Singapore: Springer Nature Singapore; 2022. pp. 669-680. (Lecture Notes in Electrical Engineering). DOI: 10.1007/978-981-19-0252-9_60

[14] Singh AK, Dwivedi ADD, Pandey A, Chakrabarti P. Design and implementation of an inverter and its application in ring oscillator circuits using an organic-thin-film-transistor based on an FTM-derived channel. Semiconductor Science and Technology. 2021;**36**(12):125006

[15] Vyas S, Dwivedi ADD, Dwivedi RD. Effect of gate dielectric on the performance of ZnO based thin film transistor. Superlattices and Microstructures. 2018;**120**:223-234

[16] Singh MK, Anusha K, Dwivedi ADD. Enhancing device characteristics of pentacene-based organic transistors through graphene integration: A simulation study and performance analysis. AIP Advances. 2024;**14**(8): 085113

[17] Dwiwedi ADD, Dwivedi RD, Dwivedi RD, Vyas S, Chakrabarti P. Numerical simulation of P3HT based organic thin film transistors (OTFTs). International Journal of Microelectronics and Digital Integrated Circuits. 2015;**1**(2):13-20

[18] Song C, Yin C, Qu H. Electronic microstructure and thermal conductivity modeling of semiconductor nanomaterials. Microelectronics Journal. 2021;**108**:104988

[19] Chandra V, Sinha N, Mathur G. Modeling, numerical simulation and performance optimization of P3HT:PC $_{70}$ BM based bulk hetero junction organic solar cells. Journal of Nanoelectronics and Optoelectronics. 2022;**17**(4):579-587

[20] Dadhich S, Garima Mathur ADD, Dwivedi. Numerical simulation, electrostatic and physical compact modeling of C8-BTBT-C8 organic thin film transistor. IJNeaM. 2024;**17**(2): 222-236

[21] Zannoni A. On the quantization of the monoatomic ideal gas [Internet]. arXiv; 1999. Available from: http://arxiv.org/abs/cond-mat/9912229

[22] Mohammad SN. Fermi energy and Fermi-Dirac integrals for zincblende-symmetry narrow-gap semiconductors with spherical energy bands. Journal of Physics C: Solid State Physics. 1980; **13**(14):2685

[23] Bouhassoune M, SLM M, Van Bobbert PA, Coehoorn R. Carrier-density and field-dependent charge-carrier mobility in organic semiconductors with correlated Gaussian disorder. Organic Electronics. 2009;**10**(3):437-445

[24] Bronstein H, Nielsen CB, Schroeder BC, McCulloch I. The role of chemical design in the performance of organic semiconductors. Nature Reviews Chemistry. 2020;**4**(2):66-77

[25] Hack M, Shaw JG, LeComber PG, Willums M. Numerical simulations of amorphous and polycrystalline silicon thin-film transistors. Japanese Journal of Applied Physics. 1990;**29**(12A):L2360

[26] Kemp M, Meunier M, Tannous CG. Simulation of the amorphous silicon static induction transistor. Solid-State Electronics. 1989;**32**(2):149-157

[27] Belykh SF, Palitsin VV, Adriaens A, Adams F. Effect of the relaxation of the electron subsystem excitation in metals

on the ionization probability of sputtered atoms. Nuclear Instruments and Methods in Physics Research Section B: Beam Interactions with Materials and Atoms. 2003;**203**:172-177

[28] Kastalsky AA, Shur MS. Conductance of small semiconductor devices. Solid State Communications. 1981;**39**(6):715-718

[29] Cuevas A. The recombination parameter J0. Energy Procedia. 2014;**55**: 53-62

[30] Gueorguiev VK, TzE I, Dimitriadis CA, Popova LI, Andreev SK. Electron trapping probabilities in hydrogen ion implanted silicon dioxide films thermally grown on polycrystalline silicon. Microelectronics Journal. 2000; **31**(3):207-211

[31] Shockley W, Read WT. Statistics of the recombinations of holes and electrons. Physics Review. 1952;**87**(5): 835-842

[32] Salleo A. Charge transport in polymeric transistors. Materials Today. 2007;**10**(3):38-45

[33] Grifoni M, Hänggi P. Driven quantum tunneling. Physics Reports. 1998;**304**(5):229-354

[34] Cordes H, Baranovskii SD, Kohary K, Thomas P, Yamasaki S, Hensel F, et al. One-dimensional hopping transport in disordered organic solids. I. Analytic calculations. Phys Rev B. 2001;**63**(9):094201

[35] Arkhipov VI, Heremans P, Emelianova EV, Adriaenssens GJ, Bässler H. Charge carrier mobility in doped semiconducting polymers. Applied Physics Letters. 2003;**82**(19): 3245-3247

[36] Dadhich S, Dwivedi ADD, Mathur G. Surface charge based modeling of TIPS-Pentacene TFT. In: 2021 8th International Conference on Smart Computing and Communications (ICSCC) [Internet]. Kochi, Kerala, India: IEEE; 2021. pp. 303-307. Available from: https://ieeexplore.ieee.org/document/9528257/

[37] Fabrication. Numerical simulation and compact modeling of Ph-BTBT-C10 organic thin film transistor. JoPC. 2024; **12**(5):1-18

[38] Bilgaiyan A, Cho SI, Abiko M, Watanabe K, Mizukami M. Flexible, high mobility short-channel organic thin film transistors and logic circuits based on 4H–21DNTT. Scientific Reports. 2021;**11**(1):11710

[39] Tripathi AK, Ashish MYN. Mobility with negative coefficient in Poole–Frenkel field dependence in conjugated polymers: Role of injected hot electrons. Organic Electronics. 2010;**11**(11): 1753-1758

[40] Singh TB, Meghdadi F, Günes S, Marjanovic N, Horowitz G, Lang P, et al. High-performance ambipolar pentacene organic field-effect transistors on poly (vinyl alcohol) organic gate dielectric. Advanced Materials. 2005;**17**(19): 2315-2320

Chapter 5

Comparative Study of DNTT-Based Low-Voltage BGBC, BGTC, TGBC and TGTC Configurations of OTFTs

Kadiyam Anusha and Arun Dev Dhar Dwivedi

Abstract

In this chapter, we present the numerical simulation and performance analysis of dinaphtho[2,3-b:2',3'-f]thieno[3,2-b]thiophene (DNTT)-based OTFTs by solving fundamental semiconductor equations using finite element method. This work offers a device simulation-based comparative study of low-voltage dinaphtho[2,3-b:2',3'-f] thieno[3,2-b]thiophene (DNTT) based on BGBC, BGTC, TGBC, and TGTC configurations of organic thin film transistors (OTFTs). Technology Computer Aided Design (TCAD) has been performed using ATLAS device simulator from Silvaco Inc. by taking into account the density of defects states (DOS) and field-dependent mobility model. A number of electrical metrics, including mobility, threshold voltage, current on-off ratio, transconductance, and subthreshold slope, have been extracted from experimental data as well as from the data obtained from the TCAD simulation of the device's I-V characteristics. For the bottom gate top contact (BGTC) configuration of DNTT-based OTFT, TCAD simulation results are found to be in good agreement with the experimental results reported by others. Additionally, we have retrieved the device performance parameters by running TCAD simulations for the top gate top contact (TGTC), bottom gate bottom contact (BGBC), and top gate bottom contact (TGBC) configurations of the OTFTs. Electrical performance parameters of all four configurations of the OTFTs have been compared and tabulated.

Keywords: OTFTs, TCAD simulation, Modeling, dinaphtho[2,3-b:2′,3′-f]thieno[3,2-b]thiophene(DNTT), BGBC, BGTC, TGBC and TGTC configurations

1. Introduction

The prospective uses of organic thin-film transistors (OTFTs), including low-cost radio frequency identification systems, low-end smart cards and most notably, flat panel display driving, have garnered a lot of research interest in the area of OTFTs worldwide [1–4]. The conducting organic substance, first reported by Heeger, Diarmid, and Skirakawa [5] in 1977, is the foundation of organic electronics. Together, they received the 2000 Noble Prize for this innovative technology [5]. The use of organic semiconductors in electronics has created a great deal of opportunity for the creation of innovative applications, including printed, flexible, organic and plastic electronics [6, 7]. Organic electronics is needed for future technological items including electronic skins, flexible circuits and sensors and displays because of its low cost, biocompatibility, great flexibility and lower temperature manufacturing. For flexible and wearable electronic applications, organic thin-film transistors (OTFTs) have emerged as the most practical transistor technology. Because of their inherent flexibility, low processing temperature, biocompatibility, lightweight, and low cost, organic thin-film transistors (TFTs) have drawn a lot of interest for use in next-generation electronics applications, including flexible electronics, wearable/stretchable devices, electronic skin and monolithic 3D ICs.

Over the past few decades, DNTT (p-type), one of the many organic semiconductors (OSC) that are currently on the market, has been investigated for use as an active channel material in OTFT manufacturing. Even when produced at ambient temperature, DNTT exhibits significantly high field-dependent mobility, low threshold voltage and great stability. DNTT has been utilized to realize organic devices including OLEDs and OTFTs because of its exceptional flexibility and uniformity [8, 9]. For device characterization, organic transistor threshold voltage (V_{TH}) and field-dependent mobility (μ_{FE}) are crucial because they indicate the onset of a significant charge accumulation in the channel and define the transistors' operating voltage [9, 10].

For the experimental study, DNTT is generally used because of its high ionization potential, which makes it air-stable. Compared to other organic semiconductors, DNTT is more resistant to heat and chemical degradation. Recent research has shown that DNTT, DPh-DNTT and C10-DNTT are viable p-type organic channel materials for the fabrication of flexible OTFTs [11–47].

High performance OTFTs are developed using a variety of designs, newly developed p/n-type channel materials, and substrate/dielectric/electrode materials. Because of their high threshold voltage, OTFTs have a rather high working voltage, which is the main barrier to their commercialization. Lowering the operating voltage of OTFTs requires high permittivity dielectrics [12–47].

OTFTs are used in various applications that require coverage of large area, flexibility in structure and very low cost, which cannot be made with crystalline silicon. Manufacturing OTFTs at very lower temperatures enables a large range of substrate possibilities and builds organic material-based transistors as future candidates for various low-cost electronic applications that require substrates which is flexible such as electronic paper, RFID applications, smart cards, touch screen mobile phones and display made by active matrix panel. Innovative human minds, however, soon sought a special type of organic or polymer semiconductor-based TFT as the active layer material that presents astonishing possibilities on substrates of flexible plastic, providing the world with low-cost, thin and printable future technology, flexible electronics, rugged and lightweight displays. Organic transistor-based circuits may be useful in applications where high speed is not necessary. Semiconductors made by organic materials are actually a new type of materials that contain polymers and very small molecules, which shows semiconducting properties. OTFTs are employed in a variety of applications that cannot be produced using crystalline silicon and that call for very cheap cost, flexible structure and wide area coverage. Producing OTFTs at extremely low temperatures opens up a wide range of substrate options and positions organic material-based transistors as potential future candidates for a number of low-cost electronic applications that need flexible substrates, including RFID applications, smart cards, touch screens, mobile phones and active matrix display screens.

However, quick thinking led to the search for a unique kind of organic or polymer semiconductor-based TFT as the active layer material, which offers incredible potential on flexible plastic substrates and will enable low-cost.

In this work, we present a comparative study of low-voltage dinaphtho[2,3-b:2′,3′-f]thieno[3,2-b]thiophene (DNTT) based on BGBC, BGTC, TGBC and TGTC configurations of organic thin-film transistors (OTFTs). TCAD simulation results were contrasted and compared with the device's experimentally measured results for BGTC configurations of OTFTS. Section II describes the simulation and device structure. Section III provides an overview of the main conclusions that were reached based on the investigations we carried out.

2. Device simulation for BGBC, BGTC, TGBC and TGTC configurations of OTFTs

This section presents the method of extracting parameters from the simulated model and through experiments reported by others. It also introduces the device structure under consideration, explains ATLAS simulation based on finite elements, and compares TCAD simulated results with experimentally measured results. This section also covers the impact of valance band edge intercept density (NTD) on the electrical characteristics of the OTFT.

2.1 Device structure and TCAD simulation

We present a comparative study of low-voltage dinaphtho[2,3-b:2′,3′-f]thieno[3,2-b]thiophene (DNTT)-based OTFTs in BGBC, BGTC, TGBC and TGTC configurations of organic thin-film transistors (OTFTs). For this purpose, we created the (DNTT)-based OTFTs in BGBC, BGTC, TGBC and TGTC configurations using ATLAS, which is a commercial device simulator from Silvaco Inc. Numerical simulations have been performed by solving the Poisson's equation, continuity equations for electrons and holes, and drift-diffusion equations for electrons and holes [22].

We have simulated all four configurations of DNTT-based OTFTs in BGBC, BGTC, TGBC and TGTC configurations, as shown in **Figure 1**.

Initially, we looked at the configuration of the bottom gate top contact (BGTC). Here, on a flexible PEN substrate, the OTFT is intended to be the bottom gate top

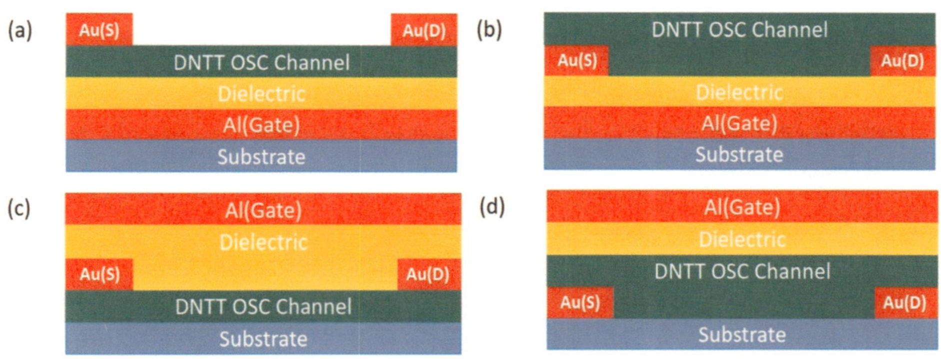

Figure 1.
Schematic of OTFTs (a), BGTC (b), BGBC (c) and TGTC (d) TGBC configurations of OTFTs.

contact. This structure's schematic, which is depicted in **Figure 1(a)** [25], is comparable to one that was investigated experimentally by Zaki et al.

It is believed that a gate dielectric is made up of a self-assembled monolayer (SAM) of n-tetradecylphosphonic acid, measuring 1.7 nm in thickness, and an aluminum oxide layer, measuring 3.6 nm. Next, an 11 nm-thick organic semiconductor layer of DNTT was applied to the AlOx/SAM gate dielectric. In order to provide a high capacitance per unit area and enable the transistors and circuits to function at a low voltage of roughly 3 V, it is anticipated that the AlOx/SAM gate dielectric (5.3 nm) has a very tiny thickness. The OTFT features a 200 μm channel length and a 400 μm channel width.

In addition, we designed three other structural configurations, which have similar dimensional parameters to the bottom gate top contact (BGTC) structure (**Figure 1 (a)**), namely, bottom gate bottom contact (BGBC) (**Figure 1(b)**), top gate top contact (TGTC) (**Figure 1(c)**), and top gate bottom contact (TGBC) (**Figure 1(d)**). **Table 1** lists the dimensions of the device, and **Table 2** lists the simulation parameters utilized in the ATLAS simulation.

We initially created the physical structure and device dimensions, including the position of the electrical connections comparable to those published in Ref. [25], to compare reported experimental results with simulated results. **Figure 1(a)** [25] shows a cross-sectional view of DNTT-based OTFT in BGTC configuration.

Based on physics, the ATLAS simulation model necessitates the simultaneous solution of basic device equations, including the defect density of states equation, drift-diffusion carrier transport equation, electron and hole continuity equation, and Poisson's equation [22–47]. The physical properties of organic semiconductors are determined by polaron mobility and synthesis [23, 26].

Eq. (1) provides the distribution of fixed charges and the internal mobility of the carriers, which are used by the Poisson equation [23, 27, 28] to compute the electric field strength in the given device:

$$\nabla.E = \frac{\rho}{\in} \tag{1}$$

where $\in$ the permittivity of the region, and ρ is the volume density of charge given by

$$\rho = q[n - p - N_D{}^{+} + N_A{}^{-}] \tag{2}$$

Parameter	Dimension
Dielectric thickness of AlOx (d_{ox})	3.6 nm
Dielectric thickness of SAM (d_{SAM})	1.7 nm
Semiconductor thickness of DNTT (d)	11 nm
Channel length (L)	200 μm
Channel width (W)	400 μm
Thickness of gold for source and drain contact	25 nm
Thickness of aluminum	30 nm
Gate to contact overlap (L_{ov})	10 μm

Table 1.
Device dimension for ATLAS simulation.

Material simulation parameters	Value of parameters
Intrinsic p-type doping in DNTT (N_A)	2×10^{17} cm^{-3}
N_{TA}	4×10^{16} cm^{-3} eV^{-1}
N_{TD}	1×10^{16} cm^{-3} eV^{-1}
W_{TA}	0.4 eV
W_{TD}	0.9 eV
W_{GA}	0.2 eV
W_{GD}	0.2 eV
E_{GA}	0.9 eV
Electron mobility	7×10^{-4} cm^2/V-s
Hole mobility	2.1 cm^2/V-s
Pool Frankel factor (betap.pfmob)	7.758×10^{-4} eV(V/cm)$^{1/2}$
ΔE_a is the zero-field activation energy	1.792×10^{-3} eV

Table 2.
Material parameters for ATLAS simulation of the OTFT.

where the electron density is denoted by n, the hole density by p, the ionized donor density by N_D^+, and the ionized acceptor density by N_A^-. Poisson's equations are amended by including an extra term Q_T, which stands for the trapped charge given in (3), in order to account for it:

$$\rho = q(n - p - N_D^+ + N_A^-) - Q_T \tag{3}$$

Q_T is equal to q(pT - nT). In this case, p_T and n_T represent the ionization densities of the donor-like and accepter-like traps, respectively, where p_T = density $\times$ f_{tD} and nT = density $\times$ f_{tA}, where the likelihood of ionizing accepter-like traps and donor-like traps, respectively, is represented by f_{tA} and f_{tD}.

The characteristics of organic TFTs are determined by the overall density of defect states (DOS), or *g(E)*, which takes into account the effects of four bands: two deep energy bands (one donor-like and the other acceptor-like), and two tail bands (corresponding to the donor valence band and acceptor class conduction band). The Gaussian distribution, as provided by the following equation, is used to model them [23, 29–47]:

$$g(E) = g_{TA}(E) + g_{TD}(E) + g_{GA}(E) + g_{GD}(E) \tag{4}$$

where

$$g_{TA}(E) = N_{TA} \exp\left[\frac{E - E_c}{W_{TA}}\right] \tag{5}$$

$$g_{TD}(E) = N_{TD} \exp\left[\frac{E - E_v}{W_{TD}}\right] \tag{6}$$

$$g_{GA}(E) = N_{GA} \exp\left[-\left[\frac{E_{GA} - E}{W_{GA}}\right]^2\right] \tag{7}$$

$$g_{GD}(E) = N_{GD}\ \exp\left[-\left[\frac{E - E_{GD}}{W_{GD}}\right]^2\right] \quad (8)$$

Here, the subscripts (T, G, A, D) stand for the tail, Gaussian (depth), acceptor, and donor states, respectively. E is the trap energy, E_C is the conduction band energy, and E_V is the valance band energy. N_{TA} and N_{TD} and W_{TA} and W_{TD} of exponential tails states are used to characterize DOS. With a Gaussian distribution, N_{GA} and N_{GD}, W_{GA}, and W_{GD} and E_{GA} and E_{GD} of DOS are used to characterize it.

In this ATLAS simulation, the Poole-Frenkel mobility model is utilized to explain the hopping-based charge transport process in organic materials such as DNTT. The Poole-Frenkel model explains why the greater excitation of trapped carriers by an electric field leads to the conduction of charge carriers. Due to the high hole mobility of DNTT, the Poole-Frenkel mobility model [32] is applied, which can be represented mathematically as follows:

$$\mu(E) = \mu_0\ \exp\left[-\frac{\Delta E_a}{kT} + \left(\frac{\beta}{kT} - \gamma\right)\sqrt{E}\right. \quad (9)$$

The electric field is represented by E, the Boltzmann constant by k, the temperature is designated by T, the Poole-Frankel factor by β, fitting parameter is γ, the zero field activation energy by ΔE_a, the field-dependent mobility by μ(E), and the zero field mobility by μ_0, Atlas simulations included thermoionic emission and Poole-Frankel barrier lowering.

2.2 Comparison of TCAD simulated results with experimentally measured results

The experimentally measured characteristics of DNTT-based organic thin-film transistors in bottom gate top contact (BGTC) arrangement are displayed in **Figure 2**, together with the transfer characteristics derived from the TCAD simulation [25]. By ramping the gate to source voltage (VGS) from 0 to -3 V while maintaining a constant drain voltage of -2 V, the transfer characteristics are obtained. The experimental transfer properties of the manufactured device reported by others and the simulated transfer characteristics coincide extremely well [25].

The output characteristics of the DNTT-based organic thin-film transistor in BGTC configuration are shown in **Figure 3**, along with the output characteristics of the DNTT-based OTFT in BGTC configuration that were measured experimentally [25].

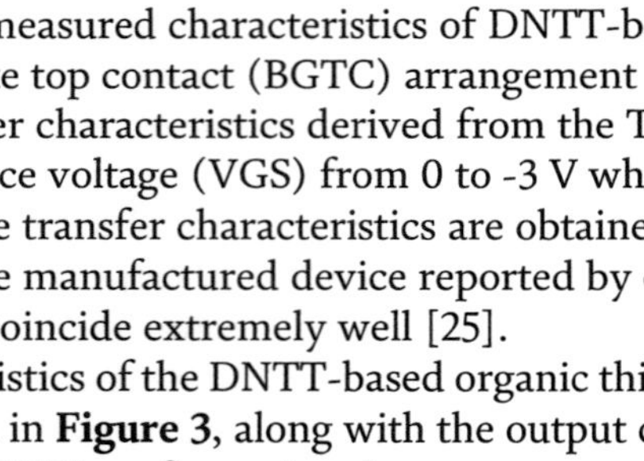

Figure 2.
I_{DS}-V_{GS} characteristics of DNTT-based OTFT in BGTC configuration: Comparision of TCAD simulation and Experimental I_{DS}-V_{GS} characteristics.

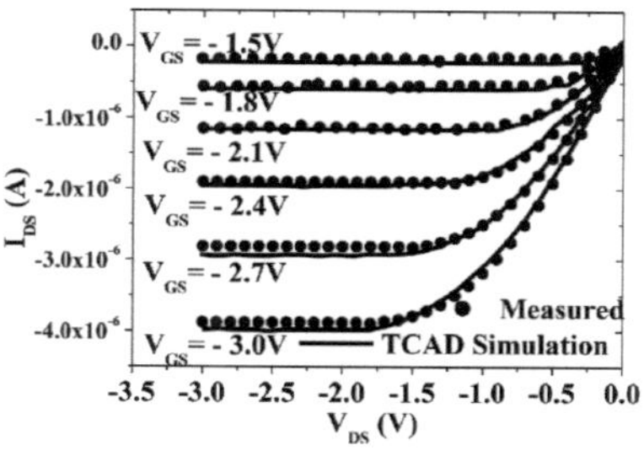

Figure 3.
I_{DS}-V_{DS} characteristics of DNTT-based OTFT in BGTC configuration: Comparision of TCAD simulation and Experimental I_{DS}-V_{DS} characteristics.

The gate to source voltage (VGS) is held constant at −1.5, −1.8, −2.1, −2.4, −2.7 and − 3.0 V while the drain to source voltage (VDS) is varied from 0 to -3 V to produce the desired output characteristics. The computed results are validated since the simulated and experimental output properties of the constructed device nearly match.

The transfer curve, as shown in **Figure 2**, provides some more electrical parameters to be determined as given below [33].

1. *Threshold voltage (V_T)*: Corresponds to gate to source voltage (V_{GS}) that accumulates large amounts of charge near the dielectric/semiconductor interface. One possible way to determine this parameter is to use the linear extrapolation of I_{DS}-V_{GS} at low drain to source voltage (V_{DS}).

2. *On–off ratio*: This parameter is the ratio of maximum value of drain to source current (I_{DS}) to minimum I_{DS}. It is well known that drive capability will be better if higher "on" currents, while lower "off" currents result in lower leakage currents. Therefore, a higher ratio of on–off current is preferred.

3. *Sub-threshold Slope (S_{sth})*: This parameter represents the V_{GS} required to increase I_{DS} by one decade, as seen in the sub-threshold region. It is defined as V/decade. A smaller S is preferable, resulting in lower power consumption and higher speed:

$$S_{sth} = \left(\frac{dlog(I_{DS})}{dV_{GS}} \bigg|_{max} \right)^{-1} \tag{10}$$

4. *Mobility:* Regarding mobility (μ), it directly affects the maximum I_{DS} and switching speed. There are different ways to determine mobility. Effective mobility (μ_{eff}) is considered to be the most accurate estimate, and this includes the V_{GS} effect. If g_{DS} is the drain to source conductance, W is the width of the channel, L is the length of the channel, and V_T is the threshold voltage, then the effective value of mobility can be expressed as:

$$\mu_{eff} = \frac{g_{DS}}{C_i \frac{W}{L} (V_{GS} - V_T)} \tag{11}$$

The field-effect mobility (μ_{FE}) is one of the most commonly used parameters in OTFT. It is obtained by using the transconductance (g_m) at low-voltage and dielectric capacitance as

$$\mu_{FE} = \frac{g_m}{C_i \frac{W}{L} V_{DS}} \tag{12}$$

Threshold voltage, sub-threshold swing, on/off current ratio, and effective charge carrier mobility are the main electrical characteristics that characterize OTFT performance. The speed at which charge carriers can pass through semiconductor material is described by the field-effect mobility, which is proportional to the semiconductor conductivity. They are frequently used as data to compare the performance of different organic semiconductor materials because this is directly responsible for determining transistor performance measures like maximum operating frequency, gain, etc. In over-the-field transistor transistors, VT stands for switch-on voltage or the gate voltage that significantly raises the drain-source current.

Once the TCAD simulation model is verified against experimental data, it is utilized to investigate the electrical behavior of other configurations of the OTFT, for example, BGBC, TGTC and TGBC. The transfer and output characteristics of OTFTs have been analyzed using proper physics. Transfer characteristics and output characteristics are examined for BGTC, BGBC, TGTC and TGBC OTFTs, and the simulated results are shown in **Figure 4(a)-(d)**, respectively. Performance parameters and output characteristics are also obtained by TCAD simulation. **Figure 5(a)-(d)** show

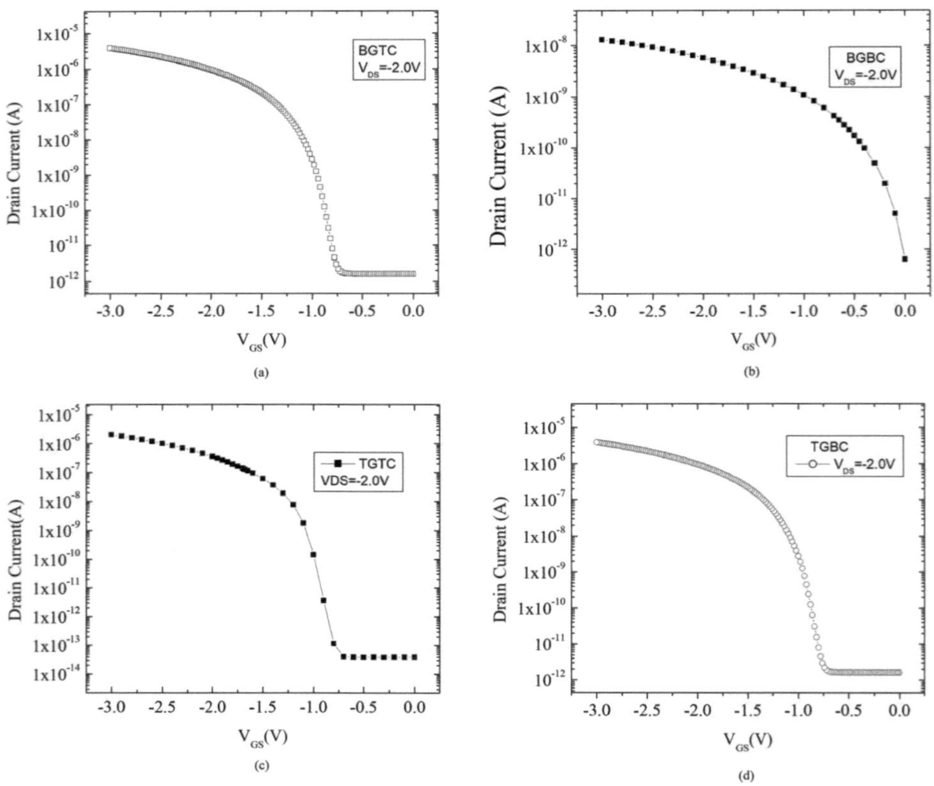

Figure 4.
Transfer characteristics of different types of OTFTs (a) BGTC, (b) BGBC, (c) TGTC and (d) TGBC.

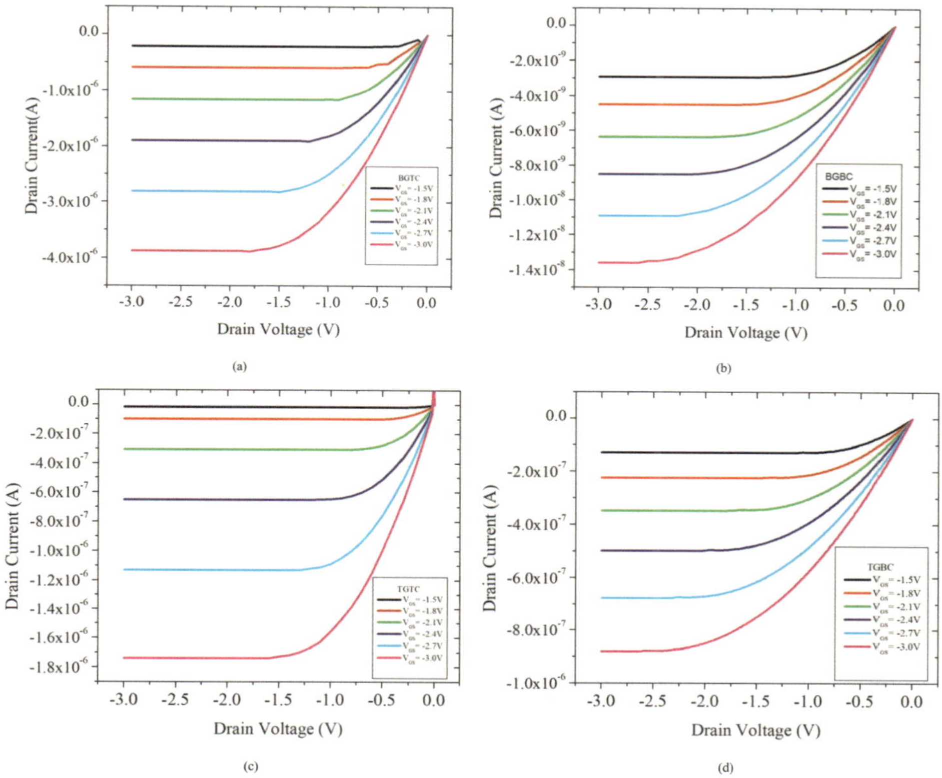

Figure 5.
Output characteristics of different types of OTFTs (a) BGTC, (b) BGBC, (c) TGTC and (d) TGBC.

the simulated output characteristics for BGTC, BGBC, TGTC and TGBC configurations of OTFT, respectively.

The same nature of characteristics is noticed for top contact (TC) and bottom contact (BC) structures, but with some reduction in the maximum current of BC structure. This is because fewer holes are injected from the source to the active layer of the organic semiconductor. TC structure exhibits more current due to the larger number of holes in semiconductors that are injectioned into it. So, the impact of the voltage difference between the Fermi energy level of metal (gold) and the HOMO level of organic semiconductor (DNTT) is more dominant due to a lesser contact area than that of the top contact structure. The key electrical parameters that define OTFT performance include effective charge carrier mobility, threshold voltage, subthreshold swing, and on/off current ratio for BGTC, BGBC, TGTC and TGBC configurations are listed in **Table 3**.

As seen from **Figure 5(a)**-**(d)**, the maximum drain current (I_{dmax}) at drain voltage of −3.0 V and gate voltage of −3.0 V, respectively, of the bottom gate top contact device is higher than that of other structures of OTFTs for same structural parameters. This means that the bottom gate top contact provides more current in the device than any other structure without any structural or electrical behavioral differences in the semiconductor-metal contact barrier. So, we can conclude that the difference in the performance of the bottom contact and top contact device is due to structural differences in the device.

Type of Structure	Sub-threshold slope (mv/decade)	Threshold voltage (volt)	On–off ratio (I_{on}/I_{off})	Transconductance (μA/V)	Mobility (cm^2 /VS)
BGTC	64.6345	−0.97 V	2.3919×10^6	3.77663	1.683
BGBC	112.119	−0.26 V	1.9×10^4	0.00750	0.00258
TGTC	62.87	−1.125 V	5.3×10^7	2.33143	1.17
TGBC	151.67	−0.8947 V	3.1×10^4	0.83	0.3711

Table 3.
Comparison of performance parameters of BGTC, BGBC, TGTC and TGBC configurations of DNTT based OTFTs.

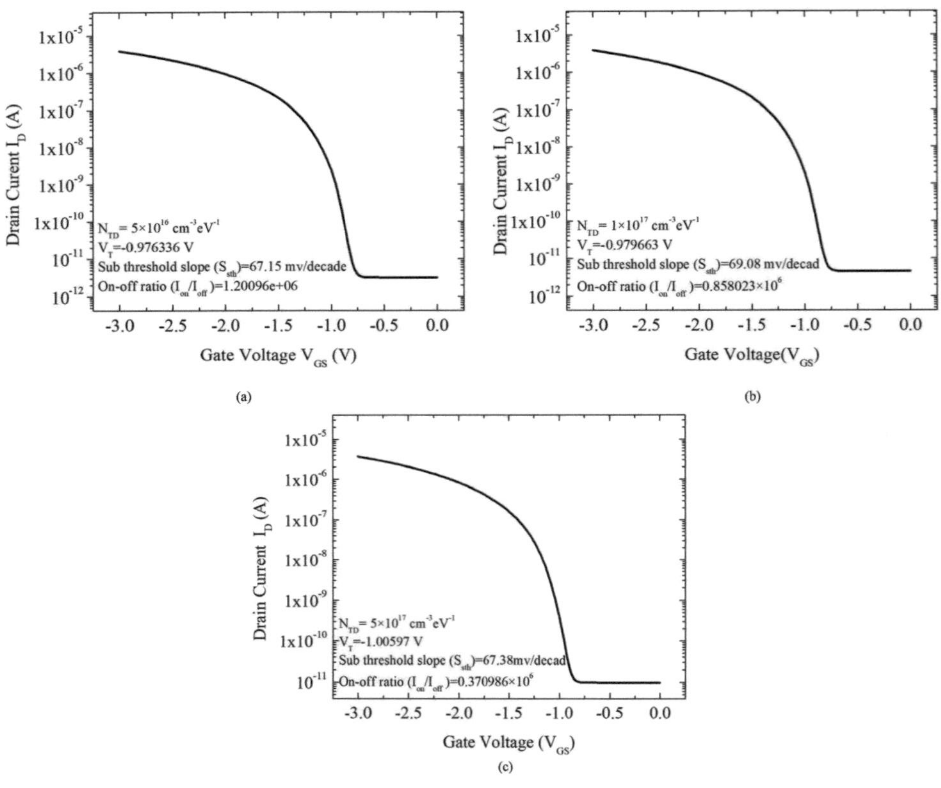

Figure 6.
Transfer characteristics of the simulated device at different values of valence band edge intercept density.

2.3 Effect of valence band edge intercept densities on electrical parameter of device

For an exponential tail distribution, as shown in equations [4–16], N_{TD} specifies the density of donor-like states in the tail distribution at the valence band edge. For an exponential tail distribution, the DOS is described by its conduction and valence band edge intercept densities (N_{TA} and N_{TD}). In p-type organic semiconductor valence band edge intercept density (N_{TD}) affects the various electrical parameters like threshold voltage, sub-threshold slope and on–off ratio. **Figure 6(a-c)** shows I_D-V_G

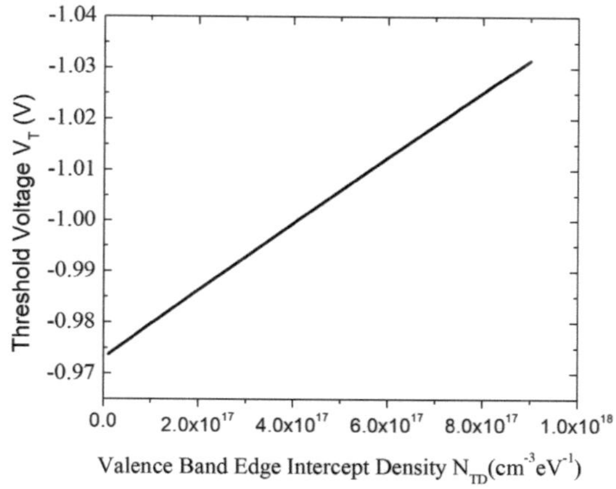

Figure 7.
Plot between valance band edge intercept density and threshold voltage.

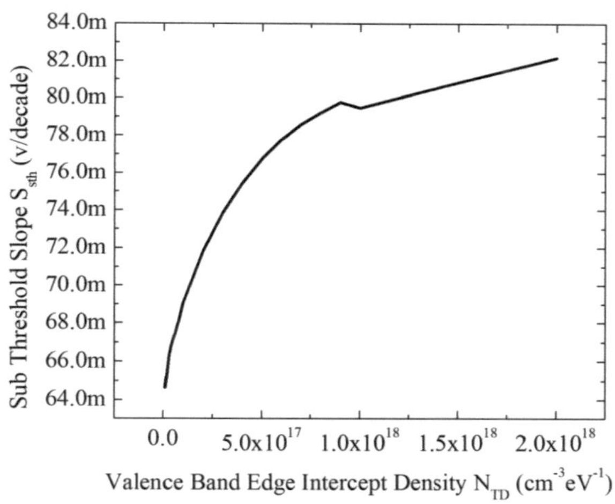

Figure 8.
Plot between valance band edge intercept density and sub-threshold slope.

transfer characteristics with various electrical parameter extractions for different values of N_{TD}. **Figure 7** shows the effect of valence band edge intercept density (N_{TD}) on threshold voltage. The graph shows that the threshold voltage increases linearly along the negative axis with an increase in valence band edge intercept density (N_{TD}). In **Figure 8**, the plot between valence band edge intercept density (N_{TD}) on the sub-threshold slope is given. It is seen from **Figure 8** that the sub-threshold slope increases with an increase in N_{TD} in the beginning, and then it saturates for higher N_{TD}.

3. Conclusions

This study presents the comparative device modeling of low-voltage DNTT-based OTFTs in BGBC, BGTC, TGBC and TGTC configurations utilizing the finite element

Performance parameter	Simulated	Experimental [25]
Threshold voltage (V_T)	−0.97 V	−1.0 V
On–off ratio (I_{on}/I_{off})	2.3919×10^6	10^6
Transconductance (g_m)	3.77663 μA/V	4.5 μA/V
Mobility (μ_{sat})	1.683 cm^2/VS	2.0 cm^2/VS
Sub-threshold slope (S_{sth})	64.6345 mv/decade	62.3mv/decade

Table 4.
Comparison of extracted parameters derived from the TCAD simulation and through experimental data for DNTT-based OTFT in BGTC configuration.

method (FEM) and a commercial device simulator called ATLAS. The model takes into account the density of defective states and the field-dependent mobility model. We performed numerical simulations of DNTT-based OTFTs by solving the fundamental semiconductor equations at every grid point of the device. Initially, electrical performance parameters of the device in BGTC configurations of OTFTs were extracted from the experimental results and from the results obtained from the TCAD simulation of the DNTT-based OTFTs in BGTC configurations and found to be in good agreement. Finally, TCAD simulations were performed for other configurations of the OTFTs, such as BGBC, TGTC and TGBC configurations. **Table 4** presents the results of a comparison of the extracted electrical parameters for the DNTT-based OTFT that were obtained from the simulated model and through experimental data for the case of BGTC configuration of OTFTs. **Table 3** shows the comparisons of the extracted electrical performance parameters for all four configurations of OTFTs, for example, BGTC, BGBC, TGTC and TGBC configurations.

Acknowledgements

The authors are thankful to SERB, DST Government of India, for the financial support under Early Career Research Award (ECRA) for Project No. ECR/2017/000179.

Author details

Kadiyam Anusha and Arun Dev Dhar Dwivedi*
School of Electronics Engineering, Vellore Institute of Technology, Vellore, Tamil Nadu, India

*Address all correspondence to: adddwivedi@fmail.com; arundevdhar.dwivedi@vit.ac.in

References

[1] Mizukami M, Hirohata N, Iseki T, Ohtawara K, Tada T, Yagyu S, et al. Flexible AM OLED panel driven by bottom-contact OTFTs. IEEE Electron Device Letters. 2006;**27**:249-251

[2] Takamiya M, Sekitani T, Kato Y, Kawaguchi H, Someya T, Sakurai T. An organic FET SRAM with back gate to increase static noise margin and its application to braille sheet display. IEEE Journal of Solid-State Circuits. 2007;**42**: 93-100

[3] Cantatore E, Geuns TCT, Gelinck GH, Veenendaal EV, Gruijthuijsen AFA, Schrijnemakers L, et al. A 13.56 MHz RFID system based on organic transponders. IEEE Journal of Solid-State Circuits. 2007;**42**:84-92

[4] Briand D, Oprea J, Courbat NB. Making environmental sensors on plastic foils. Materials Today. 2011;**14**: 416-423

[5] Shirakawa H, Louis EJ, MacDiarmid AG, Chiang CK, Heeger AJ. Synthesis of electrically conducting organic polymers: Halogen derivatives of polyacetylene, (CH) x. Journal of the Chemical Society, Chemical Communications. 1977;**16**:578. DOI: 10.1039/c39770000578

[6] Yamada M, Takeda Y, Tokito S, Matsui H. Printed organic transistors and complementary ring oscillators operatable at 200 mV. Applied Physics Express. 2024;**17**(1):011010. DOI: 10.35848/1882-0786/ad1db5

[7] Ajayan J, Sreejith S, Manikandan M, Bharath Sreenivasulu V, Aruna Kumari N, Ravindran A. An intensive study on organic thin film transistors (OTFTs) for future flexible/wearable electronics applications. Micro and Nanostructures. 2024;**187**:207766. DOI: 10.1016/j.micrna.2024.207766

[8] Liu K, Ouyang B, Guo X, Guo Y, Liu Y. Advances in flexible organic field-effect transistors and their applications for flexible electronics. npj Flexible Electronics. 2022;**6**(1):1-19. Article number: 1. DOI: 10.1038/s41528-022-00133-3

[9] Nirosha R, Agarwal R. Characterization and modeling of threshold voltage for organic and amorphous thin-film transistors. Microelectronics Reliability. 2023;**147**:115054. DOI: 10.1016/j.microrel.2023.115054

[10] Yusof NS, Mohamed MFP, Ghazali NA, Khan MFAJ, Shaari S, Mohtar MN. Evolution of solution-based organic thin-film transistor for healthcare monitoring– From device to circuit integration: A review. Alexandria Engineering Journal. 2022;**61**(12): 11405-11431. DOI: 10.1016/j.aej.2022.05.013

[11] Hiroshiba N, Kawano Y, Ongko R, Matsubara R, Kubono A, Kojima H, Koike K. Early-stage growth process of dinaphtho[2,3-b:2′,3′-f]thieno[3,2-b]thiophene (DNTT) thin film. Physica Status Solidi (a). 2023;**220**(24). DOI: 10.1002/pssa.202300252

[12] Kloes A, Leise J, Pruefer J, Nikolaou A, Iñíguez B, Gneiting T, et al. THM-OTFT: A complete physics-based verilog-a compact model for short-channel organic thin-film transistors. IEEE Journal of the Electron Devices Society. 2023; **11**:734-743. DOI: 10.1109/jeds.2023.3294598

[13] Lodha A, Singh R. Prospects of manufacturing organic semiconductor-based integrated circuits. IEEE

Transactions on Semiconductor Manufacturing. 2001;**14**(3): 281-296

[14] Petti L et al. Metal oxide semiconductor thin-film transistors for flexible electronics. Applied Physics Reviews. 2016;**3**(2):021303

[15] Yamamoto T, Takimiya K. Facile synthesis of highly π-extended heteroarenes,dinaphtho[2,3-b:2′,3′f] chalcogenopheno[3,2-b] chalcogenophenes, and their application to field-effect transistors. Journal of the American Chemical Society. 2007; **129**(8):22242225

[16] Jeon J, Murmann B, Bao Z. Fully inkjet-printed short-channel organic thin-film transistors and inverter arrays on flexible substrates. IEEE Electron Device Letters. 2010;**31**:1488-1490

[17] Lee BH, Bazan GC, Heeger AJ. Doping-induced carrier density modulation in polymer field-effect transistors. Advanced Materials. 2016; **28**(1):57-62

[18] Gundlach DJ, Zhou L, Nichols JA, Jackson TN. An experimental study of contact effects in organic thin film transistors. Journal of Applied Physics. 2006;**100**:024509

[19] Street RA, Salleo A. Contact effects in polymer transistors. Applied Physics Letters. 2002;**81**:2887

[20] Yaglioglu B, Agostinelli T, Cain P, Mijalković S, Nejim A. Parameter extraction and evaluation of UOTFT model for organic thin-film transistor circuit design. Journal of Display Technology. 2013;**9**(11):890-894

[21] Estrada M, Cerdeira A, Puigdollers J, Reséndiz L, Pallares J, Marsal LF, et al. Accurate modeling and parameter extraction methodfor organic TFTs. Solid-State Electronics. 2005;**49**: 1009-1016

[22] Buonomo A, di Bello C. On solving Poisson's equation in two-dimensional semiconductor devices. Electronics Letters. 1984;**20**(4):156-158

[23] Dwivedi ADD, Jain SK, Dwivedi RD, Dadhich S. Numerical simulation and compact modeling of low voltage pentacene based OTFTs. Journal of Science: Advanced Materials and Devices. 2019;**4**(4):561-567

[24] Fjeldly A, Ytterdal T, Shur M. Introduction to Device Modeling and Circuit Simulation. New York, NY, USA: Wiley; 1998

[25] Zaki T, Scheinert S, Hörselmann I, Rödel R, Letzkus F, Richter H, et al. Accurate capacitance modeling and characterization of organic thin-film transistors. IEEE Transactions on Electron Devices. 2014;**61**:98-104

[26] Hertel D, Bässler H. Photoconduction in amorphous organic solids. ChemPhysChem. 2008;**9**(5): 666-688

[27] van Roosbroeck W. Theory of the flow of electrons and holes in germanium and other semiconductors. The Bell System Technical Journal. 1950; **29**:560-607

[28] Juengel A. Drift-diffusion equations. Vol. 5. Chap: Springer; 2009. pp. 99-127

[29] SILVACO. ATLAS User's Manual - Device Simulation Software. Santa Clara, CA, USA: Silvaco International Inc.; 2010

[30] Vyas S, Dwivedi ADD, Dwivedi RD. Effect of gate dielectric on the performance of ZnO based thin film transistor. Superlattices and Microstructures. 2018;**120**:223-234

[31] Dwivedi ADD, Dwivedi RD, Dwivedi RD, Vyas S, Chakrabarti P. Numerical simulation of P3HT based organic thin film transistors (OTFTs). International Journal of Microelectronics and Digital Integrated Circuits. 2015; **1**(2):13-20

[32] Shim CH, Maruoka F, Hattori R. Structural analysis on organic thin-film transistor with device simulation. IEEE Transactions on Electron Devices. 2010; **57**:195-200

[33] Dwivedi ADD. Numerical simulation and spice modeling of organic thin film transistors (OTFTs). International Journal of Advanced Applied Physics Research. 2014;**1**(2):14-21

[34] Iniguez B, Picos R, Veksler D, Koudymov A, Shur MS, Ytterdal T, et al. Universal compact model for long- and short-channel thin-film transistors. Solid-State Electronics. 2008;**52**:400-405

[35] Estrada M, Cerdeira A, Puigdollers J, Resendiz L, Pallares J, Marsal LF, et al. Accurate modeling and parameter extraction for organic TFTs. Solid-State Electronics. 2005;**49**: 1009-1016

[36] UTMOST IV Spice Models Manual. Santa Clara, CA, USA: Silvaco International; 2018

[37] Kumari P, Dwivedi ADD. Modeling and simulation of Pentacene based organic thin film transistors with organic gate dielectrics. Journal of Microelectronics and Solid State Devices. 2017;**4**:13-18

[38] Dwivedi ADD, Kumari P. TCAD simulation and performance analysis of single and dual gate OTFTs. Surface Review letters. 2019;**1950145**:1-7. DOI: 10.1142/S0218625X19501452

[39] Dadhich S, Dwivedi ADD, Singh AK. Fabrication, characterization, numerical simulation and compact modeling of P3HT based organic thin film transistors. Journal of Semiconductors. 2021;**42** (7,074102):1-13. DOI: 10.1088/1674-4926/42/7/074102

[40] Singh AK, Dwivedi ADD, Pandey A, Chakrabarti P. Design and implementation of an inverter and its application in ring oscillator circuits using an organic-thin-film-transistor based on an FTM-derived channel. Semiconductor Science and Technology. 2021;**36**:125006. DOI: 10.1088/1361-6641/ac274f

[41] Dwivedi ADD, Dwivedi RD, Dwivedi RD, Zhao Q. Technology computer aided design (TCAD) based simulation and compact modeling of organic thin film transistors (OTFTs) for circuit simulation. International Journal of Advanced Applied Physics Research. 2019;**6**:1-5

[42] Dwivedi ADD, Jain SK, Dwivedi RD, et al. Numerical simulation and compact modeling of thin film transistors for future flexible electronics. In: Hybrid Nanomater: Flexible Electronics Materials. UK: IntechOpen; 2020. DOI: 10.5772/intechopen.90301

[43] Dadhich S, Mathur G, Dwivedi ADD. Numerical simulation, electrostatic and physical compact Modeling of C8-BTBT-C8 organic thin film transistor. International Journal Of Nanoelectronics And Materials. 2024; **17**(2):222-236

[44] Jain SK, Joshi AM, Dwivedi ADD. Technology and modeling of DNTT organic thin-film transistors. In: Advanced Technologies for Next Generation Integrated Circuits. UK: IET; 2020. pp. 197-211. DOI: 10.1049/PBCS049E

[45] Dadhich S, Mathur G, Dwivedi ADD. Mobile charge carrier based Modeling of 4H-21 DNTT and structure analysis of OTFT. In: 2021 IEEE 2nd International Conference on Technology, Engineering, Management for Societal Impact Using Marketing, Entrepreneurship and Talent, TEMSMET 2021. USA: IEEE Explore; 2021. DOI: 10.1109/TEMSMET53515.2021.9768714

[46] Dadhich S, Dwivedi ADD, Mathur G. Surface charge based modeling of TIPS-pentacene TFT. In: 2021 8th International Conference on Smart Computing and Communications: Artificial Intelligence, AI Driven Applications for a Smart World, ICSCC 2021. USA: IEEE Explore; 2021. pp. 303-307. DOI: 10.1109/ICSCC51209.2021.9528257

[47] Dadhich S, Dwivedi ADD, Mathur G. Numerical simulation and analytical modelling of C8-BTBT-C8 organic transistor and analysis of semiconductor thickness. Proceedings of ICTSES 2021 in book "Intelligent Computing Techniques for Smart Energy Systems". Springer Lecture Notes in Electrical Engineering. 2022; **862**:669-680. DOI: 10.1007/978-981-19-0252-9_60

Chapter 6

Polymers in Liquid Crystal Displays (LCDs): Recent Advances and Future Perspectives

Elham Bigdeli and Reza Najjar

Abstract

Nowadays, polymers are playing a vital role in the development of liquid crystal display (LCD) systems and increasingly attracting the attention of researchers in many related fields. Almost every day, many new reports on the synthesis and modification of the structure of liquid crystal polymers (LCPs) are being added to the literature, unveiling novel materials with superior electro-optical properties or exploring the performance of the existing materials to be used in various fields related to LCDs. In this chapter, we summarize the role of polymers in the advancement of liquid crystal displays (LCDs) and highlight the role of polymers in the world of light, as well as image industry. The first half of the chapter deals with a brief introduction about liquid crystalline compounds and their electro-optical characterization. Moreover, LCDs and their types are also discussed. In the second part, the various types of LCPs are described. Lastly, different methods for using polymers as a substrate for protecting LC compounds and increasing their electro-optical properties are explained.

Keywords: liquid crystal displays, liquid crystal polymers, polymer-dispersed liquid crystals, encapsulated liquid crystal/polymer core-shells, birefringence, polarization

1. Introduction

Nowadays, screens, mainly as liquid crystal displays (LCDs), have become an essential and unavoidable part of human life and are widely used in many types of devices and applications such as personal computers, media, smart phones, automotive, and signs [1]. Until the late twentieth century, during the development and generalization of the displays, the cathode ray tube (CRT) technology was extensively used in this industry. However, with the advent of LCDs in the world of screens, due to a series of special benefits such as high image quality and high definition, flat and ideal size, low driving voltages, and low energy consumption as compared to CRT, the day-to-day demand for these displays increased. Special properties of LCDs prompted the manufacturers to produce more developed LCTs to replace CRTs [2].

The main compound of an LCD is its liquid crystal (LC) composition. In fact, the LC is a state between liquid and crystal, which has properties such as fluidity like liquid and ordered orientation of molecules similar to that of a crystal. Among the most important, recent applications of LCs are optical imaging and recording [3]. The optical properties of LC materials can be influenced by the electric field (such as small electric voltage), which causes the orientation change of molecules in an LC layer, and this phenomenon is referred as the electro-optical effect. The operation of the LCDs is mainly based on the electro-optical effect [4]. Among the most necessary and vital parameters for the superior performance of a material to be used as an LCD active compound are dielectric and diamagnetic anisotropies, including birefringence, elastic constants, and viscosity coefficients [5]. The LC optical anisotropy is a result of inter and intra-molecular specific electronic, as well as steric interactions that lead to a long-range order in LC, which distinguishes the LC from an ordinary isotropic liquid [6]. Due to the electro-optical limitations, research on LC compounds with new and superior properties is in progress [5]. In most of the cases, because of their adequate optical anisotropy and fast response to an external electrical field, LC compounds with low-molecular-weight are preferred and used in display devices, whereas liquid crystal polymers (LCPs) as the functional polymeric materials have been preferred in engineering of flexible LCDs as they exhibit proper required properties based on their process capability and mechanical authority, such as good orientation property and LC phase structural flexibility, which allows their use in many applications [7].

Nowadays, LCPs as advanced compositions have gained great consideration because of having versatile supra-molecular structures resulting from their self-assembly and different phase transitions [8]. They are as more advanced LC polymers compared with the conventional LC compounds used in commercialized applications, and scientific research. With the advent of its day-to-day advancement, polymer science seeks to synthesize and provide newer compounds with better properties for use in specific applications. This chapter focuses on the role of polymers in the development of LCDs and conveys that the polymers using different strategies can make significant contributions in this area. In fact, the LC can be in the form of polymer or can be protected by a suitable polymer due to high sensitivity to environmental conditions. So far, most LCPs have been synthesized that include both the side-chain and main-chain LCPs. These reports are covered in this chapter. In many other works, it is reported that the LC compound because of its high sensitivity is used in a suitable polymer in the form of polymer-dispersed liquid crystal (PDLC) or protected in other more stable structures, such as core-shell structures, which are formed by suitable polymer.

The devices, including LCDs, which are fabricated by using LC compounds often and generally used for a long time, and potentially in sensitive environments, such as the environments with the risk of impact and pressure, or environments that are prone to high energy radiations, such as nuclear facilities, and in various aerospace applications. Long-term use of these systems can be limited due to environmental stresses, such as high energy radiations and high pressure on the LC compounds, which strongly affect their quality and function. The protection of the LC composition under these circumstances can be useful in order to secure and extend the life of the LCDs. Polymer matrix can be utilized to stabilize the LC compound to act as an optical switch [9], which ultimately can result in shortening of the response time of LC to light, widen the blue phases temperature range, or realize the distribution of gradient refractive index [10].

Over the past several years, polymers, because of their high spatial control in three dimensions, attract very much attention due to their notable characteristics such as very accurately tunable and excellent optical, electrical, and mechanical properties. Polymers, in the field of light and image, have revealed their high functionality. Here, first, we briefly introduce the LC materials and their variants, refer to some recently used new LC compounds. In the second part, we focus on the LCPs and the methods of using polymers in the LC compounds and also their protection. We consider some of the works, wherein LC was protected by the polymer.

2. Liquid crystal displays (LCDs)

Nevertheless, the LCs had been identified in 1888, but their applications in display technologies were realized in 1968 by the David Sarnoff Research Center. Since then, LCDs, because of enormous profits such as high image quality and high definition, flat and ideal size, low driving voltages, and low energy consumption, have become a good substitute for CRTs. Now, the devices, such as calculators and cell phones, were introduced in the electronic market, and also, LCDs are used in more devices due to their extraordinary properties. In fact, in our everyday life, LCDs have become one of the unavoidable devices for displaying information electronically [2].

As a non-emissive display, LCDs contain several layers, which have their own functionality and own way for optimizing, but in addition to more benefits of LCD screens, there are some shortcomings, such as limited viewing angle and non-saturated colors [11]. Viewing angle as an important parameter for LCDs indicate the maximum seeable angle for a screen without reducing its brightness and change in color and its quality (when watching TV directly in front of the screen (0° below)). In addition to the viewing angle, contrast setting is also an important parameter that an LCD designer should pay attention to that, which together determines the appearance of the LCD and the final appearance and quality of the product. Recently, with progress made in the LCD industry, the new systems, such as in-plane switching, fringing-field-switching (FFS), and multi-domain vertical aligned mode (MVA) [12] have been introduced, which solve many of these problems. These mainly consist of the methods, which are applied to improve the viewing angle. The controllability of the viewing angle of display device can be very fascinating functions that may provide the requirements for simultaneous demands for the sharing of information and also privacy protection. Some innovative techniques have been proposed for fabricating devices with controllable viewing angles, which are mainly based on the use of laminated and layered polymeric structures and applying varying voltages of 0–8 V across the PDLCs film [13].

As previously mentioned, choosing an appropriate LC compound as the core of an LCD display and using new methods related to how to use the LC in the LCD screens can solve many of these problems. In this chapter, some of the ways to solve the problems by using the polymers are mentioned. In general, there are three types of LCDs:

a. Transmissive LCDs, which are using backlights and display images with high quality in a dim light, but they are power-hungry, and more importantly one cannot see the screen in sunlight conditions.

b. Reflective LCDs, with no backlights and low contrast ratio, are energy-saving and readable in sunlight conditions but unreadable in dark places, so they would be better for outdoor usage.

c. A combination of the favorable features of transmissive and reflective displays has made the third type, which is known as transflective LCDs [14] that are readable in both bright and dark places and have transmissive and reflective sub-pixels; however, there is a shortcoming about light efficiency that is lower than two other types.

Reflective LCDs have two subgroups: polarizer-based and polarizer-free. The polarizer-based LCDs have essentially one or two polarizers and a mirror as a reflector [15]. Generally, the polarizers are widely used in display technologies. Wire grid polarizers (WGPs) as a group of polarizers consist of sub-wavelength periodic metal lines [16]. Recently, a variety of high-efficiency and highly compact WGPs than conventional polarizers have been designed that were used to polarize infrared or visible light [17].

In the polarizer-based display, LC modulates the incident light polarization. Their relatively high contrast ratio and use of mainstream liquid-crystal manufacturing equipment are favorable. But there is a deficiency that is about their low reflectance (about less than 20%) as the front polarizer cuts down the incident light intensity by more than 50% [18].

An active element of the metal-insulator-metal (MIM) reflective color LCD as one of the reflective LCDs was developed by replacing a backlight in transmissive LCDs with a diffusive reflector [19]. Therefore, MIM-type LCD, as a favorable candidate, can reduce energy consumption of active-matrix drive, and due to their low-temperature, fabrication process compatibility can lead to design the flexible displays with some degree of bending that recently have been highly noticed. For example, in many literature reports, examples of the plastic LCDs of transmissive and reflective types have been reported [20]. In another aspect, the absence of polarizers in polarizer-free reflective LCDs leads to higher reflectance and also makes it possible to use flexible plastic substrates, which have a non-uniform birefringence that causes non-uniformity in displays if polarizer(s) are used. Polarizer-based reflective LCDs are not compatible with the plastic substrates because of the lamination of the polarizer(s) on the outside of the display-cell substrates. Plastic substrates in polarizer-free LCDs offer advantages such as lightweight, ruggedness, and the possibility of production *via* roll-to-roll manufacturing process. In the polarizer-free reflective displays, the LCs either scatter, reflect, or absorb the incident light [18].

One of the main strategies to fabricate the energy-efficient LCD systems is to design and synthesize novel material [21] and techniques or to innovate new ways for the use of existing materials or systems. As an example, Afshanmehr et al. have designed and synthesized some LC compounds based on coumarin mesogen with pentaerythritol core [21] and some varying length alkane chain spacer groups connecting mesogen group to the pentaerythritol core and have evaluated their electro-optical performance [22]. They have confirmed that the spacer chain length has significant effect on the performance of the prepared LC compounds, such as relatively low threshold voltage. As another example, in a twisted nematic liquid crystal device (TNLCD), researchers have replaced the conventional polyimide alignment layer with an alignment layer that was doped with nano-P(VDF-TrFE) as a ferroelectric polymer nanoparticles or some inorganic nanoparticles or semiconductor nanostructures [23]. This modification has resulted to about 50% reduction in threshold (V_{thr}) and saturation voltages (V_{sat}) [24]. Furthermore, this type of LCs/polymer composite films may include advantages, such as long-life and low-power consumption [25]. The results of the Zhao's work have revealed that, by incorporation of 0.8 wt% of MgO nanoparticles into the PDLC films, V_{th} was reduced to around 7.5 V, which confirms the high potential of MgO-doped PDLC as an excellent choice for the low energy consumption LCD materials [26]. In fact, reflective devices were proposed to reduce the power consumption *via* permitting specific light to pass and reflecting otherwise absorptive one for recycle [27].

In guest-host reflective displays as another batch of LCD displays, dichroic dyes are doped into LCs. In one state, the dye absorbs light and the reflectance is low, and in another state, the dye does not absorb light and the reflectance is high. For example, in a guest-host LCD system, a reflective type LCD was employed to get many colors (for example 4096 colors) without a polarizer [28]. In light-scattering reflective displays, such as PDLC displays, light is scattered back out of the display and produces reflection. In these systems, the LC homogeneously spread in a polymeric matrix.

Recently, by setting the amount of hydrophilicity of polymer (such as using the long hydrophilic flexible chain in polyacrylate polymer) with anchoring interactions between the polymer chains and LC, the feature of PDLC displays have improved in different directions, such as reducing the driving voltage, response time, etc. [29].

In cholesteric (Ch) and holographic PDLC reflective displays, light is reflected from the liquid crystal. Although the quality of images in polarizer-based types is better than the polarizer-free displays, the polarizer-free displays can provide large-sized displays with low costs [18].

3. Liquid crystals

De Gennes and Prost have given the most constructive definition for LCs [30]. Friedrich Reinitzer, an Austrian botanist, worked on cholesteryl benzoate as a first LC in 1888. During the experiment, he observed that, as the temperature increased, first a hazy liquid appeared from a solid sample of cholesteryl benzoate, and then it became a clear liquid. In 1889, a physics professor named Otto Lehmann coined the term "liquid crystal" because of his studies and affirming that the substance has two melting points. In fact, in the LC state, similar to the solids, the orientational order of the molecules was kept, but in the same as in gases, liquids, and amorphous solids, no positional order in the molecular arrangement exists. In simple explanation, LC compounds are including a rigid section (most of the core) and a flexible section (most part of the side chain). Generally, liquid crystallinity achieved when there is a balance between the rigid cores and flexible chains [31]. Today, materials, which form LC mesophases, play an essential role in manufacturing of information displays, molecular sensors, light modulators, optical switches, and lots of other applications [32].

LC materials can be divided into three types:

- Thermotropic: In which the state of the LC depends on the temperature. LCs, which are used in LCDs, are all thermotropic.

- Thermotropic LC compounds in terms of mesomorphic behavior divided into two categories: enantiotropic and monotropic.

    - Enantiotropic: Same mesophases reversibly changed upon heating and cooling.

    - Monotropic: Mesomorphism observed upon cooling only.

- Lyotropic: In which the state of LC depends on the concentration and occurs as micellar structures aggregate through the solvent-induced orientation. It has high performance in biological systems [33] and occurs as a result of the accumulation of micellar structures in the solvent through the corresponding mesogens.

- Amphitropic: Temperature and concentration are determining factors.

Another assortment of LCs is based on their molecular shapes. For example, rod-like LCs called calamitic and disk-like LCs are known as discotic. LCs possess properties between solid and liquid and exist in three types of nematic, cholesteric, and smectic phases that are different in terms of structural order when they are in the external field. LCs that used in LCD screens are often in the nematic phase due to having special physical characteristics, such as broad range of working temperatures. However, all nematic phase LCs are not appropriate for utilization, for example, some of them do not function properly at room temperature. In LCDs, under the applied magnetic or electric field, the main fraction of the molecules arrange along a particular direction, can be manipulated in a predictable manner [34]. Nematic phase, derived from the Greek word "threads," is the simplest LC phase and refers to the thread-like defects observed in nematics. The organic molecules in the form of rod-shaped nematic phase, such as isotropic liquids, are partially fluid, and molecules have no positional order but can be arranged by an external field or other methods, such as adding some nanoparticles, like ITO NPs or carbon quantum dots. Aligned nematics are widely used in LCDs due to their specific optical properties of uniaxial crystals. The smectic phase, depending on the orientational and positional order degrees, is classified into categories, such as smectic A, B, etc. There are some comprehensive review articles about the controlling of the LC alignment with nanoparticles [35]. To compare, the molecular orientations in smectic A phase are in direction of normal layer, while in smectic C phase, they are somehow tilted toward that layer [36]. The Cholesteric LC is a kind of nematic LC, which contains a chiral center. This chirality causes some intermolecular forces, which causes to the alignment of the molecules at a slight angle to each other. One of the noteworthy matters about the cholesteric mesophase is the pitch. The pitch, p, equals the visible light wavelength and indicates the distance between the director to rotate one full turn in the helix, and a color will be reflected. Temperature is an important factor in controlling the pitch length in the cholesteric mesophase [37].

In fact, the term cholesteric refers to the cholesterol groups in LC structures. In cholesteric LCs, molecules are arranged parallel (such as nematic LC molecules) in thin layers. In the cholesteric structure, there is a slight displacement for the director of each layer compared to the adjacent layer's director, and this is the special aspect of the cholesteric structure. As the displacement of the adjacent successive layers accumulates, this results to the growth of a helical structures. The formation of helical structure consequently leads to the properties such as optical rotation, two-color circular polarization, as well as selective reflection. Cholesteric liquid crystal polymer films because of their chiral nematic networks have been used before to provide surface topologies by means of heating [38].

Various techniques, including some microscopic methods such as cryo-transmission electron microscopy, optical polarizing microscope (POM), scattering techniques such as neutron, X-ray, Fabry-Perot scattering, nuclear magnetic resonance spectroscopy (NMR) in combination with thermal methods such as differential scanning calorimetry (DSC) and miscibility studies are good examples of the techniques applied the identification of LC phases [39]. The enthalpies and temperatures for the transitions between different phases are measured by DSC, while it cannot identify the phase type. Polarizing microscope is a device, in which the light first polarized *via* passing from a filter afterwards from the sample and at last from the analyzer. By the observation of an LC compound on the slide of polarizing microscope while heating, the liquid crystal structure of the characteristic textures would be clearly visible, and also if it is cooled down again, it can also give these textures [40]. The translational and positional orders of the molecules in the LC and the type of LC phase can be identified by X-ray crystallography [41].

Some applications of various LC compounds include:

- Lyotropic LCs: Coating drugs to protect them against the digestive tract.
- LCDs.
- Holographic displays and storage based on photo-responsive liquid crystals.
- Quantum science and quantum engineering applications
- Cholesteric LCs, which are used in LC thermometers and cholesteric lasers.
- LCPs which are used in high-strength fibers.
- Optical imaging, generation of electric field as light hits photoconductor plates.
- Lyotropic LCs act as biological membranes due to their amphiphilic nature.

Recently, a series of newer LC compounds with better properties have been synthesized and have been used in LCDs, some of which are discussed below.

Closo-Borane and heteroborane LCs are a conventional versatile classes of compounds that have considerable capability to act as structural units for LCs. Because of some steric effects present in the structure of clusters, the nematic phases are preferred instead of smectic phases. Sometimes, materials can show unusual properties as a result of the effects of electronic and stereoelectronic phenomena. For example, for compounds with low birefringence, the observed refractive indices can be high, or they can be transparent to the UV wavelengths or even some intensity charge transfer transitions can be seen in them. In fact, the borane compounds can be very valuable in some formulations used for some electro-optical applications, such as LCDs, as they may help to lower the threshold voltages. The boron closocluster compounds because of their unique structural characteristics provide novel knowledge about the forces acting between molecules that can stabilize their relevant LC phases. For instance, isoelectronic, and basically isosteric compounds, but with distinctly varying polar and quadropolar characteristics, employing boranes and their carborane analogs can be synthesized. Geometric requirements are needed for the appearance of liquid crystal properties, especially in the heteroborane and closo-borane compounds, in the presence of five, six, ten, and twelve clusters with high molecular symmetry as rigid cores [42]. As previously mentioned, in these compounds, LC properties are also obtained as a result of the balance between rigid cores and flexible substituents. The refractive index and birefringence characteristic are among the very important parameters for the LC film materials. Hence, the determination of these parameters is of great importance in LCD studies [43].

Ferroelectric LCs (FLCs) are mainly related and interesting for the physicists, electrical and computer engineers, and chemists. Scientists are interested on them because of their bistable behavior and fast response, wide viewing angle, and high resolution. FLCs with negative dielectric anisotropy show high switching speed at low voltage due to spontaneous polarization (SP) owing to molecular chirality [44]. Nowadays, the low power and fast response LC devices are noteworthy for researchers and in the industry. FLC displays have a defect in comparison with conventional LCDs due to poor performance in moving picture mode and also in shock stability in relation to key technologies for T-V_{min} mode. However, FLCs are less popular than other types because of some shortcomings such as geometrical, optical, and mechanical instabilities [45].

Existence of a chiral center in an organic molecule leads to polar order in it. Mesogens of LCs are observed during the heating process of crystalline solid or the cooling process of isotropic liquid. At temperatures below the melting point of a material, monotropic mesogens, which are kinetically stable are observed [46]. In addition, the normal polar order has never been observed in any non-tilted LC phase, even with the chiral phase. Thus, the chiral nematic and chiral smectic A phases are categorized as chiral, acentric, and anisotropic families, but they are not polarized [47]. FLC materials can be divided into three groups:

- Regular ferroelectrics, in this category the dipole moments are parallel.
- Antiferroelectrics, in which the dipole moments are antiparallel.
- Ferroelectrics, with no usage have some subgroups such as V-shape smectics, and ortho-conic smectics.

Interestingly, interactions between LC compounds, such as cholesteric or and nanoparticles, are very considerable, for example, nanoparticles can significantly alter the properties of LC compounds. For example, the presence of carbon-based nanostructures, such as carbon nanotubes or graphene [48], or suitable polymers in an LC media has a favorable impact on LCs, by increasing the orientational order and inducing nonvolatile electromechanical memory effect. This factor leads to the electro-optical switching phenomenon and improves electro-optical properties [49].

Antiferroelectric (AF) LCs are like FLCs, although tips of molecules have been arranged and oriented in an opposite direction in subsequent adjacent layers. So, the layer-by-layer polarization points are in opposite directions [50]. The occurrence of AF arrangement is often attributed to the dipolar interactions of molecules in adjacent layers.

Generally, the important classes of LCs that are used in LCDs include the following:

- Twisted nematic liquid crystals (TNLCs)
- FLCs, faster switching times (~2 MHz)
- Super-twisted nematic liquid crystals (STNLCs).

4. Thermochromic LCs (TLCs)

As previously mentioned, LC compounds are divided into three categories:

- Thermochromic (TLC)
- Lyotropic
- Amphitropic

Among the three categories of LC compounds, TLCs mainly due to optical effects can be used in the design of LCDs. Owing to the high applicability of these compounds, in this section, we have briefly explained them.

In fact, TLCs because of their special properties have special importance in various industrial fields. The temperature-dependent color change is a specifically interesting and important aspect of LC compounds, especially for non-chiral organic LCs, such as non-chiral metal-containing LCs. Some of the LC applications such as thermometers, fever meters, gadgets, or warning signs on heaters are based on this property [51]. Temperature shapes the appearance of a certain phase, both in thermotropic and lyotropic LCs, but lyotropic ones are highly affected by the concentration because they consist of at least two compounds [52].

Despite the enormous progress in LC technology, TLCs provide the best explanation for the structure of various LC phases, mesophases, their interrelationship, and the nature of transitions [31]. Generally, many of the organic compounds, which have anisotropic shape, are TLC obtained by adjusting the temperature. They have also two subgroups: calamitic, which has rod-like structures, and the discotic phase that consists of molecules with disc-like structures. Smetic LCs are composed of regular layers of molecules with rod-like structures. In the nematic phase, the appearance of material is like liquid, but it keeps molecular orientation and arrangement. While the nematic phase appears as a marble-like structure, the columnar LCs are shaped like disks, so they differ from other types. Columns are compressed together to form two-dimensional arrays [53].

Step-wise melting is a process, which can be observed in enantiotropic LCs because the thermal energy is not enough to convert the solid material to disordered liquid at melting point, but with increasing the temperature, isotropic liquid is obtained [54]. Thermochromic spiropyrans with quasi-liquid-crystalline (QLC) properties are groups of TLCs, which have the possibility of conversion of spiropyrans *via* photo and thermochromical methods to the merocyanines. These materials are intermediate between super-cooled liquids and mesophases, which have many uses in various fields.

5. Electro-optical effects in LCs

Electro-optical switching as one of the most important phenomena of LCs leads to a widespread interest in research on optical display technology. Optical imaging and recording is a usage of LCs, in which a LC cell is sandwiched in two layers of photoconductor. For increasing of the conductivity, the photoconductor was exposed to a light beam, which is caused by the development of an electric field in LC, and the intensity of light determined the strength of produced electric field. The developed electric field was then transmitted by an electrode and recorded as an image [3].

Both nematic and smectic LCs can be applied in LCDs. The direction traveled by light from an LC material's layer specifies its optical properties. For the nematic LCDs, orientation of molecule's axes either in the same direction or perpendicular to the direction of applied external electric field would change the optical properties [55]. By exposing of the anisotropic particle suspensions to an external field, the following effects can be observed:

- The ordered phase reorientation (N, S, C)
- Forming a para-nematic (PN) phase
- Field-induced phase transitions

Methods of modulating light using LCs are [1] dynamic light scattering, [2] effect of guest-host, [3] change of nematic to cholesteric phase induced by field, [4] thermal effect addressed by laser, [5] reorientation of director axis induced by field, and [6] micron-sized droplets scattered light. Sometimes to strengthen the electro-optical characteristics of LCs, a series of nanoparticles are dispersed into the LC.

Blue phases (BPs), such as cubic crystal gold NPs, which were miscible in cyanobiphenyl-based LC dispersed in BP, have been studied as special patterns with a three-dimensionally ordered double twist cylinder, including orthogonal helical director twists for colloid crystals [56].

6. LCD screen types

Passive displays and active displays are two main types of LCDs. These terms refer to the circuits, which activate pixels. To address the signal of active or passive matrix of each pixel, an adequate display element is necessary, which is extremely difficult in manufacturing flexible displays [57]. Primarily, LCDs were based on segments, in which every segment had its own connection. As these systems were not economical, since then, the passive matrix display introduced as a solution to create larger LCDs. Passive matrix displays consist of a system of multiplexed time lines with a large number of pixels with separate connections. By using certain electrical components in passive matrix displays, the energy of these systems is not consumed for turning the desired pixels on or off. The LC is placed between conductive electrodes in a row/column form. By applying the required voltage, LC molecules align parallel to electrical field.

Each pixel is made up of the intersection of row and column electrodes, and by providing desired voltage scanning through the pixels, display represents images. By increasing the rows and columns, the applied voltage increases, which, in turn, leads to charging effect problem. In charging a row and column of pixels, although only a specific row and column is selected, rows and columns adjacent are also affected by the pulse. Activated pixels are dark, and pixels around are partially active (gray). Active pixels reduce the contrast and decrease the image quality, this effect is called as cross talk, which is introduced as a problem with these systems. By placing an electronic switch device in each LC pixel, the pixel charge is controlled. Totally active matrix, which consists of active elements, provides higher resolution but passive one does not [58].

Nowadays, LCD's resolution, contrast ratio, and other performances are somehow excellent except some technical challenges, such as sunlight readability and light efficiency. Thin-film-transistor liquid-crystal-displays (TFT-LCD) now are the main flat panel displays used in devices, and this is an output of about 50 years of perseverant attempts [59]. Active matrix displays, which are a class of flat-panel displays utilize a separate transistors for controlling of the charges on each cell in LC layer and also the screen is refreshed more as compared to top passive-matrix displays. This technique as a simple method leads to transfer the current to the LC in the target area. Similar to a capacitor, an active matrix monitor by using a transistor network can keep the charge for a limited time span. When the intended pixel receives a charge and as the transistors act as switching, it leads to the improvement of the quality of image over a passive matrix. Until before the next refresh occurs, the pixel remains active as the thin film transistor acts as capacitor and stores charge.

7. Polymers and LCDs

Generally, a variety of LC compounds used in LCDs include non-polymeric LCs and polymeric LCs (PLCs).

The compound 4-methoxybenzyliden-4′-butylaniline (MBBA) (**Figure 1**) was recognized in 1969 as the first room-temperature nematic LC. The major drawback of this compound is short stable temperature range, which is strongly influenced by impurities. This defect made it not commercially available on the LCDs, and research was started to find more stable LCs.

In 1973, Prof. Gray from the University of Hull (UK) found that cyanobiphenyl compounds exhibit a nematic phases LC at the room temperature. This discovery led to introduce the compound 4-pentyl-4-cyanobiphenyl (5CB) (**Figure 2**). This compound with improved characteristics, such as greater stability than MBBA, led to the production of the first available commercial LCDs [60].

The number of some famous LC components used in the LCDs include E202, E7, 4′-octyl-4-cyanobiphenyl (8CB), E8, ZLI-2293, and so on.

In the next sections, the role of polymers in the development of LCDs and their advantages will be discussed. The use of polymers in LCD include the direct use of LCP and indirect use for protection of LC compound by a suitable polymer.

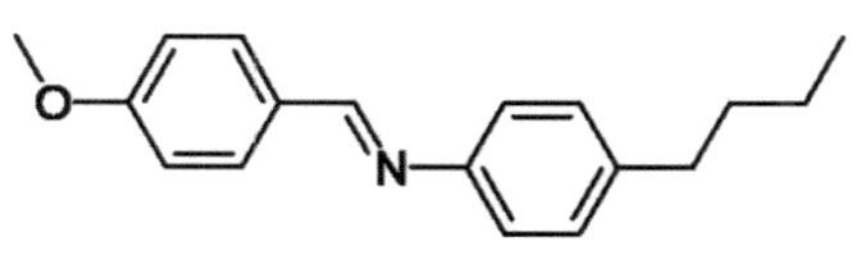

Figure 1.
The chemical N-(4-methoxy-benzylidene) -4-n-butylaniline (MBBA).

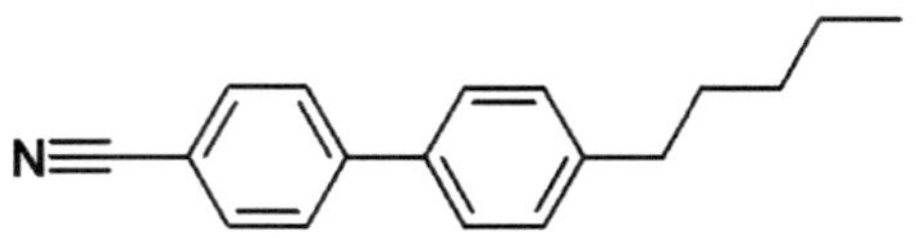

Figure 2.
The chemical structure of 4-Cyano-4′-pentylbiphenyl (5CB).

8. Liquid crystal polymers (LCPs)

As previously mentioned, thermotropic LCPs are interesting field for researchers because of their capability to self-assemble into different well-ordered structures and their interesting physical properties such as flexibility, low viscosity, low thermal expansion coefficient, very low moisture absorption, high dimensional stability, long range molecular orientation, and low cost. Moreover, other performance properties, such as modulus and stiffness values, high strength, and heat resistance that can aid in optical data storage and ferroelectric devices, nonlinear optical properties, etc. make them an ideal choice as a LCP. In many cases, blending these polymers with conventional thermoplastic materials would improve their physical and mechanical properties [61]. In main-chain LCP (MCLCP), the mesogenic groups are placed into the backbone of polymer chain, but if they are connected to the backbone by a flexible spacer then they are named side-chain LCP (SCLCP) (**Figure 3**) [62].

This anisotropy arises through coupling of the LC order of mesogens and the polymer backbone conformation. LC order and also temperature determine the degree of anisotropy, which can be modulated by UV light. Generally, in a LCP, as a thermoplastic polymer, which consist of molecular chains with regular alignment. When the LCP is heated to the LC state, the rigid moieties of the polymer chains align beside each other toward the shear flow direction and it maintains its structure even in liquid state. The monomer unit of a MCLCP consists of a rigid aromatic part (which forms a semicrystalline lattice) with flexible chains, which induce a liquid crystalline phase. SCLCPs with flexible spacers show wider LC temperature range [63].

Some applications of LCPs include: automotive parts, engineering containers, food appliances, optical devices, etc. As an inert polymer, thermotropic liquid crystalline polyester/polysulfone blends exhibit some outstanding high-temperature properties, excellent chemical resistance, and inherent flame retardancy. Also, because of its ordered structure, it exhibits high stress cracking resistance against very aggressive industrial substances. But it cannot withstand against very harsh conditions, such as high-temperature steam and concentrated boiling acids or bases [64].

LCP films with uniaxial molecular ordering have special mechanical properties related to the orientation of polymer chains, for example, such a film can tolerate fewer loads in the transversal direction than in the longitudinal direction. To overcome this problem, the two-way films, with equal longitudinal and transversal directions, were created. Although LCPs have many advantages, they suffer from some limitations. For example, it cannot be truly hermetic, it is too experimental to be used in a developing

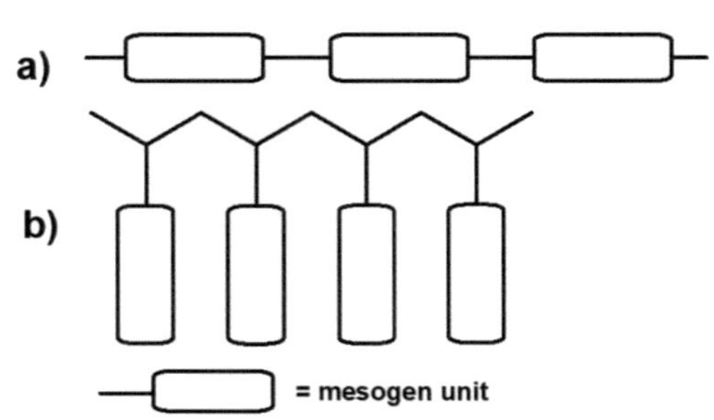

Figure 3.
Schematic representation of structure of (a) MCLCPs and (b) SCLCPs.

system, strong bases are able to dissolve them, its thermal conductivity is weak, and it has a natural surface roughness of a couple of mivcrons. As noted, the LCPs are divided into two general categories: SCLCP and MCLCP. In recent years, many groups have worked on the synthesis of LCPs and many of their defects. A noteworthy work is reported wherein electro-optical properties have been resolved and their applications in various fields reviewed. Subsequently, in this chapter, we refer to some of these works based on the types of the polymer.

8.1 Main chain liquid crystal polymers (MCLCPs)

Most of the wholly aromatic poly(azomethine) may decompose under melting temperature. In the presence of the soft aliphatic units in the polymer backbone and using of the substituent group, the melting point of these polymers would decrease [65]. TLC polyesters are noteworthy because of their mechanical strength, heat, and chemical stability. They have three sub-groups of consisting of semiflexible homopolymers, rigid homopolyesters, and copolymers, and all three groups have rigid and flexible segments in their structure [66]. Despite all shortcomings of MCLCPs, they show larger spontaneous deformations than their side-chain analogues. So far, many studies have been done on the structure-LC property relationship of LCPs. For example, the small changes in the structure, such as change in flexible spacer length, the type of mesogenic groups, etc., can affect the physicochemical properties of polymers, such as the geometry, polarity, and polarizability, and can greatly affect the LC properties of LCPs.

In the following section, to provide further insights into the structure-property relationship of LCPs, as well as to investigate the progress of polymer science in the design and preparation of new LC polymers for specific applications, some previous works about the synthesis of new LCPs are detailed, and each one of them is briefly explained.

Jannesari et al. [66] have synthesized MCLCP based on mesogenic monomers with controlled transition temperatures. They inferred that the molecular weight and the length of the polyesters are effective factors in the temperature of transition between the phases and the range of mesomorphism of the synthesized polyesters.

Grell et al. [67] have reported a new conjugated MCLCP, based on polyfluorene with octyl side chains (**Figure 4**). According to them, the polymer forms a well-defined LC state, and because of its good electroluminescence (EL) properties, it can be used in EL devices.

Xie et al. [68] have synthesized the first quasi-ladder discotic LCs bridged with metal chelates by a controlled chelation reaction under controlled experimental conditions of Cu^{2+} ions and β-diketone as ligands on the polymer chains. The resulting

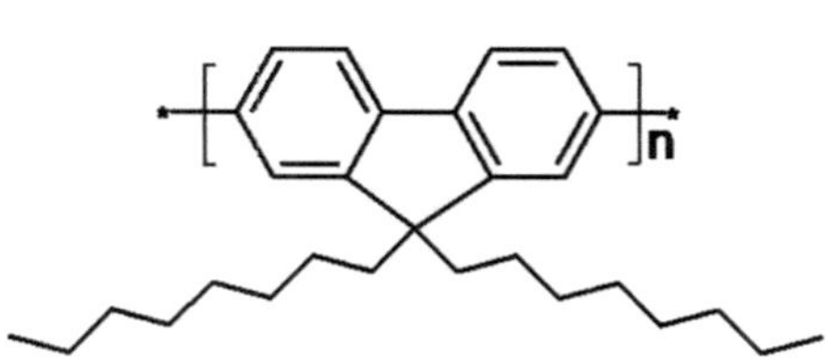

Figure 4.
Chemical structure of poly(dioctyıfluorene).

polymer showed special properties such as high metal content, good solubility, and film-forming ability.

Most of the MCLCPs are prepared using condensation polymerization method. This method leads to the preparation of polymers with high molecular weight under high-temperature reaction conditions and high vacuum. However, in the condensation polymerizations the molecular weight control and structure architecture are difficult; hence, controlled polymerization techniques are used to control the molecular weight and structure of final MCLCPs [69].

Iraqi and Wataru [70] have developed a new main-chain alkylcarbazole polymers *via* palladium-catalyzed cross-coupling polymerization route. In this reaction, the influence of alkyl group length on the carbazole monomer was investigated and the monomer with shorter alkyl group was synthesized in a faster reactions. Synthesized polymers have different molecular weights, so they exhibit different electro-optical properties.

Senthil and Kannan [71] have designed a new series of MCLC copolyesters, bearing mesogenic moieties containing ferrocenes and methyl phosphate and 2–10 methylene groups as spacers. They showed that the LC texture of polymers becomes more transparent by increasing the spacer length. Also, the presence of both parts of ferrocene and phosphorus leads to a reduction in energy-minimized structures with more molecular entanglement and also reduces the polymers melting and glass transition temperatures.

Synthesis of asymmetric MCLCPs a long ago was one of the problems in the science of LCs. Jeong et al. [72] have synthesized asymmetric bent MCLCP polyesters *via* condensation polymerization of the asymmetric A-B type monomers having varying numbers of methylene groups as spacers.

Azobenzene molecules with unique features allow molecular movement in response to the external stimulants, such as heat and light, and this factor is appropriate in photonic applications. In azo polymers, azo-containing moiety is often used to induce LC properties in the side chains. However, the number of reported works in relation to the azobenzene LCPs is low.

Toh et al. [73] have prepared polymeric and trimeric complexes of azo-based nematic LCs with H-bonding in main chain. According to their results, an increase in the spacer length of proton donors or acceptors has lowered their transition temperatures.

The azobenzene-based molecules are also used in developing of optically smart glass, which has the capability in controlling of LCs molecular alignments. A reactive azobenzene-based molecule with the ability of cis-trans photoisomerization under the UV-Vis exposure has been synthesized that allowed developing of remotecontrollable smart glasses [74].

Bagheri and Rad [75] have synthesized new series of LC polyesters, including biphenylene mesogenic group, and ester and ether linkages with different lengths as the flexible spacers. They showed that the type of linkage group can affect the thermal properties of LCPs.

Walba et al. [76] carried out catalytic synthesis of a MCLCP *via* acyclic diene metathesis polymerization using a monomer bearing dimethylvinylsilyl group in one end and a C=C double bond at the other end. The synthesized LCP product was exhibited a special and interesting type of phase transition (Iso-de VriesSmA*-SmC*- glass phase transition), which has gained a lot of interest in LC research. In this mesophase, the molecules are tilted from their natural layer with a length scale smaller than the wavelength of light, and this transition appears to be the SmA phase.

However, this type of transition is more likely to produce a helical structure in SmC phase, which permits the formation of a helical SmC director field.

Yang et al. [77] have synthesized micrometer-sized pillars made of MCLC elastomers by photopolymerization, using a thiol-ene approach. Liquid crystalline elastomers previously were prepared by polycondensation reactions. Synthesized micrometer-sized MCLC elastomers undergo an ultra-large and reversible contraction of around 400% when heated at a temperature in the range of transitions of nematic isotropic phases.

Ohkawa et al. [78] have done polycondensation reaction of bulky pyrimidine containing monomers by using LC as solvent. So, prepared polymers formed nematic LCs and exhibited higher degree of coplanarity compared to the polymers obtained using toluene as solvent. In fact, under the influence of LC and due process of aggregation polymers could be created bearing more consistency between main chains. Also, their results proved the possibility of polymerization of bulky monomers using LC solvents.

Amela-Cortes et al. [79] have synthesized polysiloxane-based, MCLC elastomers containing unsymmetrical mesogenic units. They showed that the structure of mesogenic monomer, spacer, and cross-linkers on the thermodynamic properties, such as glass transition (T_g) and clearing temperature (T_c), and type of LC phase is effective. In fact, they showed in their work, the structure-thermal property correlation.

MCLCPs also have been synthesized using a divinyl-terminated mesogenic monomer and tetramethyldisiloxane via the hydrosilylation poly-addition [80]. Synthesized polymers in a wide temperature range can form a particular type of mesophase called "dark conglomerate" chiral mesosphase (DC), which is made up of achiral LC molecules. DC phase is an optically isotropic fluid and consists of smectic LC layers that are locally deformed.

Kinoshita et al. [81] have reported on MCLCPs, bearing carbazole and oxadiazole moieties as the main chain mesogenic groups. They have synthesized polyesters *via* transesterification reaction. The polyesters formed a nematic phase with bipolar character and exhibited a well performance as polarized blue-emitting compounds in OLEDs.

Al-Muaikel et al. [82] have prepared a new series of TLC copoly(arylidene-ethers) *via* the conventional polycondensation reaction method. The most important feature of the synthesized polymers was that use of flexible monomers and selected conditions can lead to LC properties of side-chains and a rod-like conformation of the main chain.

Fang et al. [83] using a novel method under the mild conditions have synthesized a new reactive azobenzene (azo)-containing MCLCPs. They have designed azo-acrylate monomers with varying length flexible spacers and an amino end group and polymerized them with Michael addition polymerization. The synthesized polymers due to amine groups and formation of H-bonds of chains have exhibited special properties, such as higher thermal stability, lower T_g, a wide range of smectic CLC phase temperature, and reversible photo-responsive action.

Samui et al. [84] have synthesized LC siloxaneazo polymers through hydrosilylation reaction between diallylazo monomer and siloxanedihydride. Considering the fast cis-trans interconversion and also fast segments mobility, they attempted to prepare polymers with low molecular weights and use them to engine photoactuators. They have worked on siloxane backbone and obtained low molecular weight polymers. Consequently, they observed liquid crystallinity at room temperature.

Saed [85] have photopolymerized corresponding monomers to MCLC elastomers (MCLCEs) *via* novel two-stage thiol-acrylate Michael addition reaction. They have discussed more positive effects of the Michael addition-photopolymerization method and the special properties of polymers synthesized by this method. Also, by using this method, they achieved a number of thermomechanical properties and actuation performance. In addition to the above, another primary advantage of the MCLCPs is that they can be synthesized *via* chain polymerization from conventional SCLCPs [69].

8.2 Side-chain liquid crystal polymers (SCLCPs)

Because of their lower phase transition temperatures, SCLCPs are better than MCLCPs. A flexible polymethylene group as a mesogenic group can lower the melting and clear transition temperatures and make it clear to determine the order to disorder process [86]. Also, to overcome some of the deficiencies in thermotropic MCLCPs, such as high melting temperature and low solubility, many studies have been done through structural engineering and placement of mesogenic groups on the side chains [87]. The results obtained in this field have revealed that, by changing of the mesogenic groups on the polymer backbone by using spacer groups, LC compounds with improved properties can be obtained [22]. With the provided descriptions, in the past years, SCLCPs based on acrylates, methacrylates, itaconates, and polydimethylsiloxanes have received much attention [88, 89].

Generally, the chemical structure and substituent position in the mesogenic unit is an important factor that influence the characteristic properties of SCLCPs, *viz*. the rigidity of polymer backbone, tacticity photochemical and photoelectronic properties, nonlinear optical properties, and LC behavior [87].

Terminally (end-on) SCLCPs [90] and laterally (side-on) SCLCPs [77] are two kinds of SCLCPs according to the linkage type of mesogenic groups to the backbone of polymer. Presence of a flexible spacer is very important in them, but if there are bulky side groups around the backbone, there is no need to long mesogens [91].

There are a variety of methods for the synthesis of SCLCPs with various LC phases. Among the simplest methods is the free-radical polymerization of vinyl monomers. In principle, by using this method, polymers with different backbones, such as acrylates or polysiloxanes as thermotropic SCLCPs, can be synthesized [92, 93].

One problem is that, though free-radical polymerization is a simple method, one cannot control the length of the side chain and the polydispersity. As examples of living radical polymerizations, novel polymerization techniques, such as atom transfer radical polymerization (ATRP) [94] and reversible addition-fragmentation chain transfer (RAFT) polymerization [95], were introduced. Using these methods, one can have the ability to control the side-chain length and realizing low polydispersity (M_w/M_n) [96].

There are other methods, such as ring-opening polymerization and condensation polymerization, for the synthesis of SCLCPs. Condensation polymerization has advantages, such as easy selectivity over the structure and length of main chain usability for synthesizing of chiral LCs [97].

However, parameters, such as synthetic methods, structure and activity of monomers, and reaction conditions, can affect thermotropic and LC properties of the SCLCPs. In this section, some previous research on the synthesis of side-chain LCPs and innovation in each is presented.

Crivello et al. [88] have synthesized SCLCPs with polystyrene and alpha polystyrene derivative backbones, containing azobenzene mesogenic groups. In this work, the effect of various spacer lengths on the LC properties was studied.

Komiya and Schrock [98] have synthesized norbornene derivative SCLCPs bearing ((4′-cyano-4-biphenyl)yl)oxy mesogens bonded to norbornene *via* 3–7 or 9–12 methylene groups and an ester polymerized by metal-catalyzed living ring-opening polymerization. They have investigated the influence of the spacer length on the thermal and LC characteristics of the synthesized polymers. It was inferred that the type of LC phase is also closely related to the spacer length, and in exception of $n = 3$, for all of the polymers, an enantiotropic nematic phase was observed. The polymers with higher spacer length exhibited lower isotropization temperature. Cowling [99] has prepared two types of SCLCPs by cationic ring-opening polymerization (with different mesogenic cores, 2-fluorophenyl (series 1) and (3-fluorophenyl moiety (series 2)). The thermal and LC characteristics of the polymers were compared together. Thermal behavior of the 3-fluorophenyl monomers was similar to that of the non-fluorinated parent, while the 2-fluoro monomer due to structural effects showed enough lower phase transition temperatures.

Andreozzi et al. [89] have studied the enthalpy relaxation of the glasses of two SCLCP polymethacrylates containing the azobenzene mesogenic unit in its side chain with physical experiments, such as DSC. The thermograms, thus, obtained were compared with each other for monomers and random copolymer synthesized from the monomeric unit of the azo-SCLC polymethacrylate. The results showed that the polymer backbone fragility and side chain length can affect the thermal properties such as T_m, phase transition temperatures, and T_g.

Some novel mesogen-jacketed LCPs were prepared *via* free-radical polymerization [69]. In this work, the length of the polymer chains was studied with thermal and LC properties. The results showed that isotropic-LC transition depended on the length of alkoxy chains on the phenyl rings in an odd-even trend and the transition temperatures decreased with increasing this parameter. Also, the T_g temperatures of polymers were first decreased with increasing the lengths of tail and has reached a plateau at n values over 7. The synthesized polymers exhibited mesophase state at the temperatures higher than T_g.

Sato and Mizoi [100] have synthesized TLCPs containing copolyisophthalate derivatives such as nitro and methoxyazobenzene units and various types of side chain aliphatic spacers and biphenyl group in the main chain segment *via* melt polycondensation method. They showed that the variation of side chain length and consequently the polymer structure, their LC, and optical properties can be changed. Han research group [101] have prepared some SCLCP polyepichlorohydrin-based (PECH) polymers *via* nucleophilic substitution as a simple and well-known method with various mesogenic groups. The presence of the CH_2Cl groups in the backbone of polyepichlorohydrin (PECH) may accelerate the nucleophilic substitution of Cl groups by alkoxy. Regarding the great benefits of PECH, a lot of research work has been carried out on them. Later on, in 2006, this research group studied the optical and LC properties of a series of polyepicholorohydrin, polyacrylate, poly(methyl acrylate), and polystyrene-based SCLCPs using a Monte Carlo simulation method. In the presented work, some of the simulated SCLCPs were synthesized by chemical modification or polymerization. They showed that it is possible to investigate the properties such as structural, orientational, and LC properties of polymers by molecular simulation methods. The presented method provided detailed quantitative information about the investigated copolymers [102].

Verploegen et al. [103] have synthesized SCLCPs through polymerizations of polystyrene and substituted polystyrenes and binding of LC mesogens onto the polystyrene backbone. They show that the difference in the density of LC moieties bound to the polymer backbone affects the morphology of self-assembled structures formed by synthesized SCLCP copolymers. The LC character of the synthesized polymers can be monitored by the linear time dependence of the LC properties, and this leads to the strengthening of the LC content that can lead to stronger preferences. In fact, they have introduced a practical method to bind LC moieties covalently to the backbone of a siloxane-based block copolymer as a thin film, which can act as morphological controller and can be tailored for specific applications.

The preparation of SCLCPs based on polystyrene bearing various terminal groups as side chains has been done by Kang research group [104]. They have examined effect of side group type on the surface energy values and the LC properties. They have found that the types of side groups have affected the structure of synthesized SCLCPs. Because of anisotropic interactions of LC, the synthesized polymers exhibited different LC alignment actions and different LC phases. Liquid crystalline dendrimers, serving as a command surface, exhibited light-sensitive properties of alignment and topology. As a model compound, the surface anchoring, molecular conformation, and phase structure properties of azo dendrimer were explored. The density functional theory analysis and molecular dynamics have indicated that because of the slow molecular kinetics, the phase transition behavior of the model dendrimer is significantly depends on cooling rate [105]. The flexibility of backbone of the polyimide (PI) has exhibited a significant impact on the LC alignment under the exposure to the non-polarized UV light. Photosensitive PI polymers of 2,2-bis(3amino-4-hydroxyphenyl)-hexafluoropropane with varying types photosensitive moieties and also different backbone structures were studied, showing that depending on the softness of the backbone, varying irradiation times were required for the LC polymer chains to be aligned uniformly [106].

Yang et al. [87] have designed three different types of SCLCPs based on monosubstituted azobenzene moieties (ortho, para, and meta). According to the results, the transitional behavior in synthesized polymers is greatly affected by substitutions on the azobenzene ring. For example, the SCLCPs of ortho-type cannot form LC phases, but the para and meta-substituted SCLCPs exhibited smectic A phase. Also, T_g temperature of para-monosubstituted polymers is more than metha and it is more than ortho. The isotropic -LC transition temperature changes as para > meta. Interestingly, the values of enthalpies were changed inversely, and the meta-substituted compound has higher enthalpy than para isomer.

Mohammed et al. [107] have synthesized a new type of SCLCP containing mesogenic azomethine diols by condensation reactions method. They showed that when the terminal alkyl chain length increases, the T_g temperature decreases.

Park et al. [108] have synthesized new SCLCPs based on 1,4-disubstituted 1,2,3-triazole using "click chemistry." Their proposed method was a useful one for fabricating SCLCPs. The results showed dependence of thermal and electrical properties of synthesized polymers with the structure. Their results can help to synthesize some new functional materials through click chemistry method.

Petr and Hammond [109] have synthesized a new room-temperature photoresponsive SCLCP that combines azo group in the role of side chain with a low T_g temperature polysiloxane backbone and have obtained room-temperature LCs with nematic phase. The resulting polymers exhibited short response time to the light, which makes them good candidates to be used in actuator and other short response time-requiring fields.

Smokal et al. [110] have reported some SCLCPs based on methacrylic acids bearing azobenzene units and spacers with alkane structures. Further, they investigated photo-physical and photochemical properties of the synthesized polymers. Due to the special optical properties of azo group, the azobenzene compounds are studied as materials for nonlinear optical applications.

Hofmann et al. [111] have designed the SCLCPs with cholesterol-modified PEG mesogen groups and a flexible polyglycerol (PG) as side chain. They have found that the thermal and LC characteristics of synthesized polymers strongly depended on the length of the PEG chain and its crystallinity. Also, different chain lengths lead to different amphiphilicity and steric constraints in synthesized polymers.

Kondo et al. [112] have synthesized novel photoreactive LCP of cinnamic acid-activated ester. The results showed that photoreactivity, thermal, and LC stability of polymers are influenced by the type of substituent mesogen group (cinnamide and original cinnamic acid group) that relates to the intermolecular interactions and also H-bonding of polymer chains. As another way for stabilizing the LX phases, the orthoconic antiferroelectric LC mixtures doped with reactive mesogens, and UV photopolymerized for stabilization, yielded symmetric optical rise and fall times, and also their electrical response times were reduced. The spontaneous polarization in the mixtures was decreased after UV polymerization, which is the important performance of the material in electro-optical applications [113].

Ban research group [114] have used normal radical polymerization technique to synthesize some new SCLCP jacketed by the mesogenic group with two triphenylene units. The synthesized SCLCPs exhibit high T_g and form highly symmetric hexagonal columnar phase due to interactions between T_p units and main chain. Also, when the flexible spacer is abolished, the "jacketing effect" is increased that leads to overcoming the π-π interaction of T_p mesogen.

Wu et al. [115] also have worked on triphenylene-based LCs with discotic structures, but they used ATRP as a controlled polymerization method. Benefiting the advantages ATRP method, they have widely changed the length of the LC moieties in a controlled way. The polyphenylene moiety induces the disc-shaped structures, and the presence of the polyethylene leads to lower T_g temperature; flexible structures for various applications have been reported.

Xie research group [116] have prepared SCLCPs with varying length side chains bearing double bonds *via* hydrosilylation reaction. At first, they have synthesized nematic monomers that have exhibited LC characteristics at room temperature. Meanwhile, during the both of heating and cooling cycles, polymers showed an enantiotropic nematic mesophase, and this was a new initiative in the synthesis of novel LC elastomers.

Yang et al. [117] have designed combined main-chain and side-chain LCPs containing azobenzene mesogenic groups, and the results showed that the number of azobenzene groups connected to the side chains influences the LC characteristics, as well as range of transition temperatures. The synthesized polymers exhibited more smooth and wonderful LC behaviors with increasing the number of azobenzene groups.

A research article has been published by the Gimeno group [118] on the preparation of a novel type of monomethacrylate monomers, and subsequently, their ATRP polymerization to obtain SCLCPs bearing 3,4′-biphenylene group as a core unit.

Their research focused on the synthesis of these materials and the effect that structural changes and the relative positions can lead to.

Hirai et al. [119] have designed a new series of poly(methacrylate) based SCLCP bearing LC side chains *via* living anionic polymerization as a precise way to control the molecular structure of polymers. According to their findings, the short length of the chain of alkyl improved the solubility of the polymers, and consequently, this factor was also effective on the electro-optical properties of LC.

As previously mentioned, FLCPs are a very important category of LC compounds, which bear chiral mesogens in the side chain units. FLCPs are of great importance in the field as they have unique features such as low T_g and proper viscosity, interesting phase transitions, short response time, and low consumption potential.

Recently Hu et al. [120] have synthesized side chain FLCPs containing (2S, 3S)-2chloro-3-methylpentanoate. For this purpose, first, the chiral mesogenic monomers were synthesized. The results showed that chirality can be effective on the LC properties of polymers, and its effect depends on factors such as polymer backbone nature, the mesogen type and number, and the spacer type and length. They show that the different levels of chirality can be embedded to the LC molecules, and the chiral centers closer to the chain ends are more effective on chirality level. The polymer backbone nature, type of terminal group, mesogen, and spacer length are among the main parameters determining the mesomorphism of side-chain LCPs.

Supramolecular SCLCPs based on H-bonding between poly(acrylic acid) and derivatives of pyridyl Schiff base have been synthesized [121]. The investigations of the spacer length on the LC characteristics of synthesized SCLCPs have confirmed that the molar ratio of pyridyl Schiff base derivatives to poly(acrylic acid)s are very important parameters in controlling of the type of LC phase.

Polysiloxane-based SCLCPs have been prepared *via* thiol-ene click reactions, which contained substituents with or without chiral centers in the side chains [122]. The results showed that the LC and phase transition of the synthesized polymers strongly depended on the molar content of the chiral cholesteric groups. By increasing of the value of this parameter, the clearing point become very high, a very wide mesogenic temperature range was obtained, and also different types of mesophases were observed.

9. Polymer-dispersed liquid crystal (PDLC) displays

PDLCs as a special category of compounds that have both of the polymer properties, such as ability to form a film with high mechanical strength and LC properties, such as their unique electro-optical nature, with a variety of switchable electro-optical effects based on purely through a change in scattering cross-section. In recent years, the use of PDLCs in optical switches and smart windows applications extensively investigated as they can help to reduce energy consumption in buildings [123], light shutters, tunable-focus lenses, light valves, data storage, holographic films, bio [124], and many more other applications.

PDLC films have numerous advantages, including simple production of large displays, higher transmittance of light compared to conventional LCDs (which is due to the combination of LC and polymeric properties), high contrast, simple fabrication technology (no need for polarizers, aligning layers etc.), short response times, high viewing angle, high lifetime, low cost, etc. [125]. In fact, PDLCs are composed of micro/nanosized nematic phase LC particles, which are highly dispersed in a polymeric medium very uniformly [126]. The fabrication process operates based on a simple principle to prepare a polymer/LC composite film that can inhibit LCs flow and also improve mechanical deformation property of the film. When there is no external electric field, the LC particles have a random orientation, and as the refractive index of polymeric network mismatches with the LC particles refractive index, this causes to the scattering of light and consequently, the film's appearance changes to be translucent and also its color becomes milky white. By the application of electrical field, the LC molecules become oriented accordingly, so if no mismatch of refractive indexes are existed, no change in the transparency of the film will be expected [127]. Hence, having similar refractive indexes for the polymer and LC particles is one of the most important requirements for the PDLCs, in such a case the light can transmit from PDLC film without significant scattering. Also, LC should have a maximum solubility in the polymer to produce a completely homogeneous mixture [128]. As the intensity of applied electric field influences the light intensity transmitted from PDLC film, so this can be used as a base for PDLC films application as a device for measuring the strength of electrical fields [129].

Generally, various parameters are effective in PDLC films such as shape, size, and relative amount of LC particles to polymer and the refractive coefficient of both of them. Also, the polymer structure can affect the electro-optical performance of PDLC and determine the field of application of the resulting PDLC film and also their optimum working conditions, such as wide working temperature range [130]. Various types of LC compounds, such as LCs based on CN-biphenyl, are used in the preparation of PDLCs, as well as a variety of polymers such as poly(methyl methacrylate)s, poly(dimethylsiloxane) s, poly(styrene)s, epoxy resins. The polymers and LCs in different types lead to the production of PDLCs with different electro-optical properties [131].

One of the most important characteristics of the PDLCs is the driving voltage, which depends on several factors, such as the dielectric index, conductivity, the PDLC film thickness, and size of dispersed LC particles. According to the findings, the bigger LC droplets cause to the lowering of driving voltage values, as well as faster response times. In addition to the above parameters, some other physical parameters, such as structural uniformity of the film in terms of thickness and homogeneous distribution of LC particles, can also affect the PDLC characteristics and lead to a change in its applied field [132].

The LC molecules-polymer interactions greatly affect their orientations in the polymer matrix, and this is a reciprocal effect. However, polymer chains are also effective in the process of the film stretching, which leads to a change in its physical properties used in display applications [133].

There are several ways to prepare PDLC films; the most common and most convenient method is to use phase separation strategy, such as encapsulation method. This method induces a phase separation between dispersing polymer phase and dispersed LC particles by employing of a variety of methods, resulting in the formation of dispersion of the LC particles in the dispersing polymer network [134].

Three main classes for the phase separations (PIPS) can be realized, which are induced by:

a. Polymerization: As a special industrial method, a proper photoinitiator is added to start a fast polymerization *via* free-radical reaction mechanism

b. Thermally (TIPS)

c. Solvent (SIPS).

The first method offers a convenient way to control the uniformity in the size of droplets, but with this method, it is difficult to reach large size [4, 132].

In addition to the benefits of PDLCs, these systems also have a number of disadvantages. The electro-optical cells prepared based on the PDLCs mainly suffer from high cell thicknesses, bigger LC particles, and high threshold voltages, and their optical contrast ratio is low [135]. Many methods have been introduced to overcome these disadvantages, for example, by proper choosing of the system components the driving voltage can be improved. For example, De Sarkar et al. reported that using of the fluorinated monomers could significantly reduce the driving voltage [136]. It has also been reported that several ways can be used to increase the optical contrast ratio in these systems. HPDLC systems are prepared by using of polymer as dispersing medium and an adequate LC material. The anisotropic phase separation was induced by normal radical polymerization of a proper monomer.

Liu et al. [136] have reported details of the effect of surfactant to the prepolymer in these systems in order to lower the on-state voltage. They have studied the surfactants influence on performance of HPDLC Bragg gratings, and found that surfactants could significantly reduce the driving voltage.

In another work, Shim et al. [137] have examined the effect of surfactant and showed that surfactant can be effective on the diffraction index and increase this factor. Also, the surfactant with a special mechanism increases the value of spreading coefficient and results in smaller droplets, greater diffraction efficiency, reducing threshold voltage, and also faster response times.

Another way that can increase optical contrast ratio and decrease driving voltage is doping nanoparticles (NP) in the structure of PDLC films. NP-polymer influences the optical feature of PDLCs. Regardless of the few proposed routes, the exact mechanism of the action of metallic nanoparticles on the improvement of the electro-optical characteristics of PDLCs is not clarified yet. For example, researchers have improved the response time and angular properties of light transmission of PDLCs by doping with inorganic nanoparticles or metallic ones such as Ag-NPs [138], which is also has caused a significant reduction in driving voltage of the film. In another work, Tsai et al. [139] have prepared PDLC composites consisting of commercial nematic phase LC, E7, doped with various nanoclays, such as CL120 and CL42 as inorganic materials, and commercial NOA65 as dispersing polymer medium. According to their results, the nanoclay doped films have indicated a highly improved electro-optical performance compared to the LC films without doped nanoclays.

As an overview, the dynamics of PIPS method can be affected by several factors such as surface effect, pressure, monomer concentration, temperature, monomer viscosity, etc. Also, various parameters, such as LC particle size and morphology, affect the PDLCs electro-optical performances. Among them, the surface effect is one of the most important parameters. Generally, the surface effect improves LCD quality, depending on the surface anchoring strength. So that, PDLC systems consisting of high superficial surface anchoring energy and also consisted finer and more uniformly dispersed LC particles. However, a weak surface anchoring will reduce the switching voltage and increase the contrast [140].

Liu et al. [140] have prepared the flexible PDLC films showing low threshold and saturation driving voltage values by PIPS method. The results show that the rubbing of surface that affects the LC particles morphology is also experimentally and theoretically improving the optical properties. They found that nearly all of the finer LC particles are regularly arranged in the direction of surface rubbing, and the PDLC has gained a texture similar to the HPDLC. Also, they showed that the contrast ratio and the response time of surface-rubbed PDLCs were remarkably improved in comparison to the non-rubbed samples. The PDLC film thickness is another effective parameter on its electro-optical performance.

Macchione et al. [141] have reported a reverse mode PDLC system prepared by the photopolymerization of mixtures of a LC compound with low molecular mass, a negative dielectric anisotropy, and a diacrylate monomer with LC properties. The prepared PDLC film has indicated a high optical contrast, fast response times, low threshold voltages, good adhesion, and long stability. Also, their results have revealed that the orientation control and the good electro-optical properties depended on the roughness of the supports used.

Mach et al. [125] have reported a special type of the PDLCs as driven by the organic transistors, and their results showed that the displays operate robustly under bending and had a very good contrast.

Amimori et al. [133] have reported the results of a study of the opto-mechanical properties of a PDLC film that consisted of dispersions of E7 in polyvinyl alcohol polymer, which is soluble in water. They demonstrated that the two parameters, including droplet shape and polymer chains orientation, are the main influencing parameters on the final nematic director configuration.

The reflective and OLED-type LCDs require the use of polarizers in the form of films as additional components. In comparison, liquid crystal devices doped with dichroic dyes are advantageous as the need for this component is removed by replacing with switchable DLCD in OLED, resulted to the lower reflection and higher brightness in OLED systems [142]. Further, investigations on the azo dye-doped liquid crystals have revealed that the relative affinity of adsorption to and desorption of dye from the isotropic LC surface have a significant effect on its performance, and also the pumping light intensity especially at the start of the alignment of LC molecules imposed by light improves the reorientation of surface adsorbed molecules to intensify their connection to the LC surface [143].

Masutani et al. [134] have designed a fabrication technique for PDLCs doped with dichroic dyes. For this purpose, the polymer matrix was first washed that was led to the removal of LC from PDLC and the formation of porous structure. As formed porous network, subsequently, was again filled with LCs doped with dye. This resulted in optimizing the LC used in the refilling, resulting in decreased driving voltage and response times.

By using of PDLCs and non-polarized light, Ren et al. [129] have modulated the phases. Nevertheless, the change of phases is a fast process and does not dependent on the light polarization, but they showed that no significant change occur for the rest of materials phase, which can be yet proper to fabricate micro-devices, such as microlenses, as well as microprisms. They also proved that these systems have exhibited an improved threshold voltage values compared to the NP-PDLC. Polymer microprism array (PMA) used to generate a micro-optical device for beam steering, which exhibited a 0–1.65° change in beam deflection angle in response to the increase of applied voltage [144].

Bra et al. [145], using a static analytical method, have prepared the DPDLC of E7 employing supercritical carbon dioxide as a readily available, environmentally acceptable solvent that leaves no toxic residues. Considering the presence of various molecular weight ingredients in commercial E7, its fractionation was occurred in supercritical-CO_2. By increase of the temperature, the fractionation does not occur properly and a mixture of LCs with higher ratio of 5CB than E7 will be extracted into the supercritical CO_2. These findings have a great importance in the future efforts to design and develop PDLCs systems by using supercritical fluid technology. Also, when the LC particles are randomly orientated in the dispersing polymer medium a sharp decrease in the oblique angles occurs and this factor, as a weakness, increases the operating voltage.

Ren et al. [146] have demonstrated a microlens array made of NP-PDLC. For this purpose, the polymer holes that were specifically designed for microlens were filled with NP-PDLC material. They showed that the small LC droplets resulted in many benefits, for example, the corresponding devices have responded very fast, polarization-independent and no scattering were observed in them. Nanostructured-PDLCs have been employed to prepare a micro-lenticular lens array (MLA), which is operated depending on the polarized incident light. The device can exhibit a fast phase modulation and 3.8 ms of switching time between 2D/3D display device applications. The stable orientation and fast switching characteristics of the liquid crystals were improved by changing of the length of mesogenic side chains in comblike-polyimides structures. The presence of the hydrogen bonds between the chains has imparted a highly stable orientation for the mesogenic moieties in the system [147].

Polydimethylsiloxanes with various viscosities and FLC particles (ZLI-3654) were used by Sumana and Raina [148] to prepare composite films of PDFLC. They have employed SIPS method and PIPS techniques simultaneously. They also examined the performance of the PDFLC films to parameters such as temperature, polymer viscosity, alignment of LC droplets, and measuring the films performance in the states with aligned and unaligned LC molecules. According to their results, as expected, PDFLCs films have exhibited better performance in aligned state than unaligned state. The copolymers chains were interacted with the LC molecules, causing them to be vertically aligned. The copolymer of vinyl-4-heptylbenzoate grafted with vinyl-alcohol, which has amphiphilic character and a comb-like structure, was added into the LC, and the blend was coated as a thin film over the ITO electrodes. The copolymers chains were self-assembled in the interface resulting to a layer of self-assembled molecules acting as aligning layer through hydrophilic interactions of hydroxyl functionalities present on the copolymers chains. On the other hand, hydrophobic interactions of alkyl phenyl groups have caused LC molecules to the stable self-alignment of the molecules in vertical orientation, which is a critical parameter affecting device performances, such as response time and operating voltage [149].

Senyurt et al. [150] have prepared thiol-ene HPDLCs films containing NOA65 (with T_g values relatively higher than ambient temperatures) and E7 LC. It has been clarified that the electro-optical performance of the trithiol-pentaerythritoltriallylether based films, has an obvious dependence on the matrix physical nature. The prepared films had special features, such as having the fastest switching times, T_g below 0°C, and lowest switching voltages. Also, they have observed that the high concentrations of LC have led to the low switching voltages.

Kumar and Raina [151] have reported the preparation of dichroic DPDLC films containing E7 LC, polymer NOA 65, and anthraquinone-based dichroic dye. The dependence of electro-optical properties and applied voltage on dichroic dye concentration, morphology of LC droplet, and temperature was investigated.

Petrov et al. [152] have prepared PDLC films by the method of UV photopolymerization-induced phase separation *via* dispersion of variable size of E7 in a photopolymer NOA65 matrix. The PDLC films linear and quadratic performance were investigated by conventional methods, such as flexoelectric spectroscopy and laser light diffraction. According to the results, the electrically tunable notch filter performance of prepared PDLC system could find some interesting applications in modern optoelectronic devices operating in the infrasound frequency range.

He et al. [153] have prepared PDLC films *via* PIPS method using methyl acrylate monomer, a photoinitiator, and an alkanethiol as chain transfer agent (CTA). The polymer molecular mass and LC were carefully controlled and their effects on the electro-optical performance of PDLCs were evaluated. The results indicated that adding CTA resulted in a significant reduction in threshold voltage.

Wang et al. [154] have reported the preparation of specific and well-ordered structures in a PDLC film through phase separation of LC particles from polymer network assisted by a substrate that was chemically patterned, i.e. an ITO glass patterned with concentric Au ring. The effects of parameters, such as time changes relative to the size and the number of LC particles, were investigated. The results indicated that the observed phase separation has started with a significant influence of the substrate surface.

Jin and Park [155] have designed and synthesized PDLCs based on acrylamide copolymers *via* radical polymerization. They have synthesized copolymers of acrylamide with styrene, acrylamide with methyl methacrylate, and investigated together with 4-pentyl-4′-cyanobiphenyl (BL087) as a commercial LC compound. They have explained the effect of fluorine atom insertion to acrylamide-based copolymer chains and also its influence on PDLC system performance.

Ding research group [156] have prepared PDLC films through a series of acrylate-based monomers, which were curable by heat or UV light. The presence of these curable monomers was allowed to regulate the amount of curing (crosslinking), and consequently, the pour sizes in the polymer matrix, which is a determining parameter to improve the PDLC films performances.

Li and co-workers [157] reported on the fabrication of some PDLC films *via* photopolymerization of LC monomers and polymerizable photoinitiator composites. They examined the effect of the monomer structure, such as chain length and its softness on the electro-optical performance of PDLC system.

Koo research group [158] reported a type of PDLC system in the presence of a special cross-linker based on bisphenol and tris(hydroxyphenyl) ethane. Their results showed that the cross-linker compound, although were used in a very little amount, could have had a remarkable positive influence on the performance of the PDLC system. Also, when the cross-linker concentration was exceeded from a certain limit, it has affected parameters, such as driving voltage and response time.

Novel PDLC films were prepared by Yan et al. [159], which contained poly(methyl acrylate)-b-polystyrene as polymer matrix in PIPS technique. The results indicated that the addition of polystyrene homopolymer was ended to an improvement in the applied voltage; meanwhile, in the absence of polystyrene, the electro-optical quality of the system was reduced.

To be used in polymer-based optical fibers, Lallana et al. [160] have proposed a system composed of a changeable optical attenuator system combined with a multiplexer. They have used a PDLC device containing required pixels to embed a multiplexer and a variable optical attenuator bearing inputs ports equal to number of pixels. These PDLCs have required no polarized light and exhibited highly contrasted outputs, and also, their capability for grayscale imaging was high.

The influence of BL038 LC droplets in HPDLC's reflection-mode system was studied by Bacchiocchi research group [161]. They have observed that the nano droplets of LC and polymer particles are arranged in a random layer, which was led to the formation of a diffraction grating.

Nataj et al. [162] have designed HPDLC with the highest diffraction efficiency *via* a new monomer bases on urethane trimethacrylate through fabricating switchable diffraction grating method.

Natarajan et al. [163] have prepared HPDLCs using thiol-ene polymers function as better hosts for HPDLC.

Chidichimo et al. [164] have designed the reverse mode PDLC films obtained by the photopolymerization of homeotropically aligned LC solutions by varying the amounts of two different LC diacrylate monomer. Comparative results revealed that the prepared PDLC system due to its outstanding optical contrast capable of being used in the windows. They also showed that the system could be completely influenced by the effect of an appropriate electric field and switched from a high lucid into a very turbid state. The morphology of PDLC has a very important impact on its electro-optical properties.

Kashima et al. [165] have prepared the PDLC film with PIPS method from monomers/crosslinking agent/initiator/LC mixture. They have found that the chemical structure, molecular weight, and chain length of crosslinking agents directly affect the LC performance of PDLC systems.

Hinojosa and Sharma [138] have manufactured a PDLC film that was consisted of dispersions of the E44 LC and a mixture of acrylate-based monomers and also a small amount of cross-linker and photoinitiator and have added a very small amount of Au NPs into the film materials. Their investigations have revealed that the doping of PDLC films with Au NPs had a major impact on their performance, and a significant reduction in threshold voltage, increase in the optical transmission and response speed for them were observed.

Hsiao and Chang [166], *via* incorporation of chiral azobenzene molecules as LC material in a PDLC system, have designed an optically switchable transmission grating system, which was not dependent on polarization. They presented that the addition of chiral LC molecules to the system has caused the nematic LC phase to become a chiral phase, which increases the efficiency of grating system. This was led also to the improved performance of the prepared PDLC system, like no dependence of the system on polarizing light.

Lin and Tsou [167] have designed another system for modulation of LC phases employing of PDLC surface pinning effect, which was not dependent on the light polarization. The developed system was usable in a lot of devices such as laser beam steering systems, light modulators, and microprisms.

Yu et al. [168] have studied photo-responsive behaviors in PDLC-like hybrid films. They have fabricated the films using polymeric LC particles, which were lightresponsive, as dopants and used a non-photoactive polymer as a hosting material. According to their results, hydrophobic LC particles with rough morphology can have a maximum interaction with the polymer matrix. They have also observed that the prepared PDLC system can be bent to the light due to topological shape change and mesogenic alignment.

Jung et al. [169] have developed combinational structures that was consisted of guest dye-doped PDLC. Regarding its special impact, PDLCs doped with dyes has a high capability to avoid the influence of the light scattered from the PDLC surface, so it would be usable for designing of devices, such as reflective optical shutter. The higher reflective property of the fabricated system than normal PDLCs enables it to act as reflective optical shutter.

Su et al. [170] have prepared PDLC films consisted of azobenzene-derived LC (azo-LC). Moreover, they have studied the influence of concentrations of azo-LC on optical switching efficiency and response time. Their results show that the changes of the amount of LC can decrease the response time.

Parab et al. [171] have prepared PDLC film composed of PMMA and commercial LC E44. They have proved that the performance of PDLCs depended on the LC concentration, polymer-LC phase separations, and size and shape of LC particles dispersed in the film.

In lowering of the threshold voltage of PDLCs, the electrical conductance plays a remarkable role. Uniform doping of the polymer films with conductive materials, such as nano graphite can enhance its electrical conductance, which is studied by Wang et al. [172] proved that the higher concentrations of doped nano graphite have exhibited significant reduction in the operating voltages of the films. The reduction in threshold voltage of PDLCs was also achieved by doping of the films with various metal nanoparticles such as gold nanoparticles [173], functionalized Ag nanoparticles [174], Pd nanoparticles [175] metal oxide nanoparticles like magnetic γ-Fe2O3 nanoparticles [176], and metal salts of organic compounds, such as zinc acrylate nanoparticles (NPs) [177]. Zito and Pissadakis [178] have fabricated of a mixture of polymer and LC-based fiber Bragg grating (PLC-FBG). They achieved this by using visible two-beam holography, wherein, a mixture is made through photo-induced polymerization method of a prepolymer, and then the LC composition infiltrated into the holes of polymer film. The designed system has exhibited special characteristics, such as high temperature and bending sensitivity.

Parab et al. [179] have prepared PDLC films of PMMA-co-BA and commercial E44 LC compound. They have also studied some of the important parameters, such as the impedance and phase angle, dielectric coefficient, relaxation frequency, and the relaxation frequency in the system. Their results have suggested that, by controlling of the amount of LC loaded into the material, the dielectric properties of the PDLC system can be tailored.

Yu et al. [180] have prepared a fast and visible-light responsive PDLC/nanoscale graphene oxide (GO) as the photo-absorbent heat source nanocomposite films, which was composed of one of the available LC (5CB) compounds. Polyvinyl alcohol (PVA), as the photo-inert polymer host, was used as the film matrix. Under the influence of irradiation with visible light, the nanocomposite film prepared by photochemical reaction has bent toward the light source. The results revealed that the graphene oxide and its derivatives due to their specific properties, such as photothermal effect and high thermal conductivity, and due to the change in the type of phase transitions have led to the improvement of LC properties of the system.

Malik et al. [128] have prepared PDLC film composed of a polymer matrix based on poly(methyl methacrylate) and two famous LC compounds, such as E7 and E8 *via* solvent-non-solvent technique. Then, thermo-optical and also electro-optical performance of the system were studied. According to their results, film containing E7 liquid crystals has performed been better than E8 films.

Eun Jung et al. [181] have developed a method to enhance the electro-optical performance of reflective-type PDLCs doped with a dye. The method was based on the use of an acrylate-based UV-curable polymer due to its role as a separator of two PDLC films, which can improve their electro-optical performance.

Deshmukh et al. [182] have prepared PDLCs *via* PIPS method by using mixture of a prepolymer, a dye, and a nematic LC compound. Subsequently, they have confirmed that the dichroic dye has a determining effect on the LC particle size and phase separation. By controlling of the concentration of dye, a favorable electro-optical result such as minimum threshold and high contrast can be obtained.

Lu et al. [183] have prepared PDLCs via UV photopolymerization of the acrylates and have investigated the effect of some parameters such as temperature and crosslinker concentration on the electro-optical properties of resulting systems. They proved that the moderate size of LC microdroplets and dispersibility of polymer matrix have main effects on performances of PDLCs. The investigated factors, such as UV light intensity and temperature, have been proven to be influential on the morphology of polymer matrix and was led to an improvement in PDLCs performances.

Lee research group [127] have fabricated thiol-vinyl PDLCs films with improved performance by using PIPS method without the addition of photoinitiators (PIs). The PDLCs fabricated by adding PI and without addition of PI were exposed to UV light (10 days), revealing a remarkable durability for both types of films. Meanwhile, the films produced without PI exhibited reasonable physicochemical performance and the change in its haze value was negligible after UV exposure.

Marin et al. [131] have prepared PDLCs based on polyvinyl alcohol boric acid (PVAB) and 5CB as LC by a conventional encapsulation method. The performance of nematic LC (twisted) cells was improved *via* incorporation of the multi-walled carbon nanotubes (MWNTs). Also, it has been proved that the chemical modification of the surface of MWCNTs has resulted to their better performance by improving better dispersion MWNTs in the LC matrix [184]. The functionalization of the carbon nanotubes was caused to improve films driving voltage by 18.7% [185] compared to the non-functionalized case [186].

Quantum dot (QD)-based LCD system may easily suffer from elevated temperatures and water exposure by reducing its fluorescence, and this severely limits their practical application in long-term use. Hence, many attempts are on the way to overcome this weakness by encapsulation or use of especial microstructures as protective substrate for QD. In a typical work poly(styrene-methyl-methacrylate-acrylic acid) nanoparticle/polyamide 66 nanofiber film was used as loading substrate, which has exhibited high fluorescence property, which has been stable for long times, even at the temperatures of about 200°C [187].

In another work films containing QDs [188] and vinyl-ether cross-linked LCs dispersed in the polymer matrix have exhibited fluorescence enhancement and improvement in water-resistant performance [189].

Kim research group [190] have prepared and investigated the optical performance of PDLCs with NOA65 (as dispersing polymer medium) and E7 as LC in the different concentrations *via* UV irradiation method, and then the segregation of E7 LC as micro-sized particles in dispersing polymer medium was occurred.

Song research group [191] has prepared PDLC films, which contained a kind of polymer matrix based on the epoxy/acrylate hybrid. They have studied the effect of LC droplets size on the performance of PDLC films. Due to the difference in polymerization rates between the epoxy and acrylate monomers and the UV intensity gradient, a gradient in the size of LC particles dispersed in films was observed.

Khaligh et al. [192] have prepared uniform PDLC films composed of E7 as LC material, NOA-65 monomer, and SiO_2 microsphere as spacers to give thickness uniformity. They have proposed that films prepared by Ag nanowires can act as smart PDLC windows.

PDLCs with varying size of dispersed LC droplets have been used to prepare optical diffuser films. The intensity of irradiating UV light and also monomer structure has significant effect on the resulting PDLC films micro-structure. The optical diffuser films fabricated by using UV curable composites of acrylate/LCs. When spherically dispersed LC droplets in PDLC networks were ≈3.0 μm in size, exhibited haze value of 88.5%, and its transmittance was 90.0%, such film is suitable as a bottom diffuser film. Meanwhile, the films obtained by 10.0 μm dispersed LC droplets, the haze and transmittance were 39.2% and 90.2%, respectively, making it proper as a top diffuser film [193]. In another research, by embedding of micron and submicron-sized LC material droplets in cellulose acetate biopolymer matrix, and optimizing of the concentration of the dispersed LCs, ultrahigh haze films of ≈99% were obtained [194].

Light-shaping PDLC films prepared by well-designed materials with good vertical LC alignment may exhibit angle-selective scattering properties, which their responses based on the angles of the irradiating light. These films may function based on birefringence or volumetric light-shaping mechanisms, helping faster spread of backlight, resulting to the fewer mini-LEDs or very thinner display panels. It is expected that such a birefringent PDLC films will find widespread use in emerging mini-LED backlight LCDs [195].

10. Encapsulated liquid crystal/polymer core-shell (ELCPCS) microcapsules/nanocapsules

Although, the term encapsulate is used for the PDLCs, recently ELCPCS systems are designed as one of the encapsulated systems that they are more stable and coherent structures compared with the previous LC/polymer composite systems. In simple terms, a microcapsule/nanocapsule is a small sphere with a uniform wall around it, and in the encapsulation process, tiny droplets of LC are surrounded with a continuous polymer coating to give discrete capsules [196, 197]. This trapping technology is mainly used in various fields, such as to control the release rate of active agents, protect sensitive materials from harmful and hazardous environment, reduce mammalian toxicity, and sustain the performance of the active ingredients [198, 199]. In most works of this method, the aim was to protect reactive substances from the environment, and in addition to this target, this method was also used to enhance the electro-optical performance of the LCs and as a successful way of stabilizing, packaging, and protecting LC mixtures [200, 201]. The microencapsulation process has been the most versatile, widely applicable method. In addition to the normal synthetic procedures noted in the preparation of the PDLCs, another way is by coating a thin layer of polymer on the surface of LC using a suitable polymerization method, such as interfacial polymerization. Generally, the functional performances of such capsules are highly dependent on the particle morphology and the structure characteristics of the polymeric shell, such as crosslinking density, composition, thickness, porosity, and crystallinity of the shell. These factors can be controlled by some of the parameters such as material parameters and polymerization process conditions. The material parameters, such as monomer weight ratio, core-to-monomer shell ratio, the nature of the emulsifier, monomer structure, and core material composition [198], can be used for employed. Generally, the core-shell particle structures because of all of the above appetites and the shell's ability to change the core properties have received considerable attention.

11. Conclusions

Researches concerning LCDs are underway as everyday many novel applications for LCDs are being found or improvements for the current applications are demanded. In recent years, the polymer science in different ways has tried to help this progress.

During the last two decades of the millennium, that is, 2000–2018, more than one hundred new works have been addressed this issue. The synthesis of a number of new main chain or side chain LC structures has been reported, leading to generation of newer effective methods on the influence on the structure-property correlation. Also, in some cases, polymer compounds have been used to protect the LC composition and to improve its electro-optical properties. In this comprehensive review, using references to several research articles, we have made a sincere attempt to demonstrate effectively the role of polymers in the development of LCDs.

Acknowledgements

The authors would like to thank the financial supports from the Office of Postgraduate Studies of the University of Tabriz and Iranian Nano Initiative.

List of abbreviations

LCD	liquid crystal display
CRT	cathode ray tube
LC	liquid crystal
LCP	liquid crystal polymer
PDLC	polymer-dispersed liquid crystal
POM	optical polarizing microscope
DSC	differential scanning calorimetry
TFT-LCD	thin film transistor-liquid crystal display
MCLCP	main-chain liquid crystal polymer
SCLCP	side-chain liquid crystal polymer
TLC	thermochromic liquid crystal
T_g	glass transition
FLC	ferroelectric liquid crystal
ATRP	atom transfer radical polymerization
PIPS	polymerization induced phase separation
TIPS	thermally induced phase separation
SIPS	solvent-induced phase separation
NOA65	Norland optical adhesive
ITO	indium tin oxide
ELCPCS	encapsulated liquid crystal/polymer core-shell

Author details

Elham Bigdeli and Reza Najjar*
Polymer Research Laboratory, Faculty of Chemistry, University of Tabriz, Tabriz, Iran

*Address all correspondence to: najjar@tabrizu.ac.ir

References

[1] Van Krevelen DWF, Poelman R. A survey of augmented reality technologies, applications and limitations. International Journal of Virtual Reality. 2010;**9**(2):1-20

[2] Lagroix HE, Yanko MR, Spalek TM. LCDs are better: Psychophysical and photometric estimates of the temporal characteristics of CRT and LCD monitors. Attention, Perception, & Psychophysics. 2012;**74**(5):1033-1041

[3] Mohanty S. Liquid crystals — The 'fourth' phase of matter. Resonance. 2003;**8**(11):52-70

[4] Maschke U, Coqueret X, Loucheux C. Polymer-dispersed liquid crystal materials prepared by electronbeam-initiated polymerization. Journal of Applied Polymer Science. 1995;**56**(12):1547-1555

[5] Schadt M. Liquid crystal materials and liquid crystal displays. Annual Review of Materials Science. 1997;**27**(1):305-379

[6] Priest RG. Theory of the frank elastic constants of nematic liquid crystals. Physical Review A. 1973;**7**(2):720

[7] Hentschke R. Equation of state for persistent-flexible liquidcrystal polymers. Comparison with poly (γ-benzyl-L-glutamate) in dimethylformamide. Macromolecules. 1990;**23**(4):1192-1196

[8] Hammond MR, Mezzenga R. Supramolecular routes towards liquid crystalline side-chain polymers. Soft Matter. 2008;**4**(5):952-961

[9] Hirabayashi K, Wada M, Amano C. Compact optical-fiber variable attenuator arrays with polymer-network liquid crystals. Applied Optics. 2001;**40**(21):3509-3517

[10] Ren H, Fan Y-H, Wu S-T. Prism grating using polymer stabilized nematic liquid crystal. Applied Physics Letters. 2003;**82**(19):3168-3170

[11] Arias AC, MacKenzie JD, McCulloch I, Rivnay J, Salleo A. Materials and applications for large area electronics: Solution-based approaches. Chemical Reviews. 2010;**110**(1):3-24

[12] Lu R, Wu S-T, Lee SH. Reducing the color shift of a multidomain vertical alignment liquid crystal display using dual threshold voltages. Applied Physics Letters. 2008;**92**(5):051114

[13] Han C, Zhou LE, Ma H, Li C, Zhang S, Cao H, et al. Fabrication of a controllable anti-peeping device with a laminated structure of microlouver and polymer dispersed liquid crystals film. Liquid Crystals. 2019;**46**(15):2235-2244

[14] Hsieh J-L, Choi W-K. Step-shaped electrode for low-voltage and high-optical-efficiency blue-phase transflective liquid crystal displays. Liquid Crystals. 2019;**46**(7):1043-1051

[15] Kubo M, Ochi T, Narutaki Y, Shinomiya T, Ishii Y. Development of "advanced TFT-LCD" with good legibility under any ambient light intensity. Journal of the Society for Information Display. 2000;**8**(4):299-304

[16] Kim SY, Gwyther J, Manners I, Chaikin PM, Register RA. Metalcontaining block copolymer thin films yield wire grid polarizers with high aspect ratio. Advanced Materials. 2014;**26**(5):791-795

[17] Jensen MA, Nordin GP. Finiteaperture wire grid polarizers. Journal of the Optical Society of America. A, Optics, Image Science, and Vision. 2000;**17**(12):2191-2198

[18] Yang DK. Review of operating principle and performance of polarizerfree reflective liquid-crystal displays. Journal of the Society for Information Display. 2008;**16**(1):117-124

[19] Diest K, Dionne JA, Spain M, Atwater HA. Tunable color filters based on metal– insulator– metal resonators. Nano Letters. 2009;**9**(7):2579-2583

[20] Baeuerle R, Baumbach J, Lueder E, Siegordner J. A MIM driven display with colour filters on 2″ diagonal plastic substrates. In: SID Symposium Digest of Technical Papers. Vol. 30(1). 1999. pp. 14-17

[21] Afshanmehr A, Najjar R. Synthesis and characterization of four-armed Pentaerythritol- and Coumarin-based materials as potential liquid crystals. Chemistry of Natural Compounds. 2023;**59**(1):26-31

[22] Afshanmehr A, Najjar R, Safari E, Asadpour-Zeynali K. Low drivingvoltage four-armed pentaerythritol and coumarin based compounds as potential materials for LCDs. Optical Materials. 2023;**145**:114439

[23] Kumar A, Priyam MH, Prakash J, Wang L, Singh G. Recent advances on semiconducting nanomaterials–Ferroelectric liquid crystals nanocomposites. Journal of Physics: Condensed Matter. 2022;**34**(1):013004

[24] Nimmy JV, Cherumannil Karumuthil S, Varghese S. Nano P(VDF-TrFE) doped polyimide alignment layers for twisted nematic liquid crystal devices. Liquid Crystals. 2020;**47**(1):36-41

[25] Zhou L, Liu S. Development and prospect of viewing angle switchable liquid crystal devices. Crystals. 2022;**12**(10):1347-1357

[26] Zhao Y, Li J, Yu Y, Zhao Y, Guo Z, Yao R, et al. Electro-optical characteristics of polymer dispersed liquid crystal doped with MgO nanoparticles. Molecules. 2022;**27**(21):7265-7275

[27] Tsai C-C, Wu S-T. Broadband wide-angle polarization converter for LCD backlight. Applied Optics. 2008;**47**(15):2882-2887

[28] Uchida T, Wada M. Guest-host type liquid crystal displays. Molecular Crystals and Liquid Crystals. 1981;**63**(1):19-43

[29] Ahmad F, Jamil M, Woo LJ, Jeon YJ. The investigation of molecular affinity involved in poly (ethylene glycol) based polymer-dispersed liquid crystal display. Colloid and Polymer Science. 2012;**290**(7):599-606

[30] Prost J. The Physics of Liquid Crystals. 2nd ed. Oxford (UK): Oxford University Press; 1995

[31] Axenov KV, Laschat S. Thermotropic ionic liquid crystals. Materials. 2011;**4**(1):206-259

[32] Hussain A, Pina A, Roque A. Biorecognition and detection using liquid crystals. Biosensors and Bioelectronics. 2009;**25**(1):1-8

[33] Forrest BJ, Reeves LW. New lyotropic liquid crystals composed of finite nonspherical micelles. Chemical Reviews. 1981;**81**(1):1-14

[34] Demus D, Goodby JW, Gray GW, Spiess HW, Vill V. Handbook of Liquid Crystals, Low Molecular Weight Liquid Crystals I: Calamitic Liquid Crystals. Weinheim (Germany): John Wiley & Sons; 2011

[35] Malik P, Singh G, Supreet PP, Kumar A, Castagna R. Recent advances and future perspectives on nanoparticles controlled alignment of liquid crystals for displays and other photonic devices. Critical Reviews in Solid State and Materials Sciences. 2023;**48**(1):57-92

[36] Priestly E. Introduction to Liquid Crystals. New York: Springer Science & Business Media; 2012

[37] Yu A, Bobrovsky BN, Shibaev VP, Springer J. New chiral nematic materials with photovariable helical supramolecular structure for reversible optical data recording. Advanced Materials. 2000;**12**(16):1180-1181

[38] Liu D, Broer DJ. Liquid crystal polymer networks: Switchable surface topographies. Liquid Crystals Reviews. 2013;**1**(1):20-28

[39] Sagalowicz L, Michel M, Adrian M, Frossard P, Rouvet M, Watzke HJ, et al. Crystallography of dispersed liquid crystalline phases studied by cryo-transmission electron microscopy. Journal of Microscopy. 2006;**221**(2):110-121

[40] Sahara M, Yano S, Ikemoto K, Maejima K. Liquid crystalline phase transitions for 4, 4″-dialkyloxyquaterphenyl homologues. Liquid Crystals. 1993;**15**(6):929-931

[41] Clegg PS. X-ray studies of the phases and phase transitions of liquid crystals. Acta Crystallographica Section A: Foundations of Crystallography. 2005;**61**(1):112-121

[42] Ringstrand B. Boron clusters as structural elements of novel liquid crystalline materials. Liquid Crystals Today. 2013;**22**(2):22-35

[43] Kanwar A, Ghodke S, Gajghate V. Determination of refractive index and birefringence of nanoparticle-doped liquid crystals. International Journal of Thermophysics. 2023;**44**(3):35

[44] Akaogi J, Koizumi Y, Ono M, Furue H. Application of ferroelectric liquid crystals to optical devices. Molecular Crystals and Liquid Crystals. 2014;**596**(1):106-112

[45] Banabic D, Lazarescu L, Paraianu L, Ciobanu I, Nicodim I, Comsa D. Development of a new procedure for the experimental determination of the forming limit curves. CIRP Annals-Manufacturing Technology. 2013;**62**(1):255-258

[46] Ikuma N, Tamura R, Shimono S, Uchida Y, Masaki K, Yamauchi J, et al. Ferroelectric properties of paramagnetic, all-organic, chiral Nitroxyl radical liquid crystals. Advanced Materials. 2006;**18**(4):477-480

[47] Giesselmann F, Zugenmaier P, Dierking I, Lagerwall S, Stebler B, Kašpar M, et al. Smectic-a*–smectic-C* transition in a ferroelectric liquid crystal without smectic layer shrinkage. Physical Review E. 1999;**60**(1):598

[48] Kumar A, Pratap Singh D, Singh G. Recent progress and future perspectives on carbon-nanomaterialdispersed liquid crystal composites. Journal of Physics D: Applied Physics. 2022;**55**(8):083002

[49] Shcherbinin D, Konshina E, Solodkov D. The effect of CdSe/ZnS quantum dots on the rotational viscosity and charge carrier concentration of a

nematic liquid crystal. Technical Physics Letters. 2015;**41**(8):781-783

[50] Lopatina LM, Selinger JV. MaierSaupe-type theory of ferroelectric nanoparticles in nematic liquid crystals. Physical Review E. 2011;**84**(4):041703

[51] Seredyuk M, Gaspar AB, Ksenofontov V, Reiman S, Galyametdinov Y, Haase W, et al. Room temperature operational thermochromic liquid crystals. Chemistry of Materials. 2006;**18**(10):2513-2519

[52] Burducea G. Lyotropic liquid crystals I. Specific structures. Romanian Reports in Physics. 2004;**56**(1):66-86

[53] Shimizu Y, Kurobe A, Monobe H, Terasawa N, Kiyohara K, Uchida K. Calamitic and discotic mesophases formed by kinetically controlled rod–disc alternation of molecular shape in a triphenylene–azobenzene mesogenic system. Chemical Communications. 2003;**14**:1676-1677

[54] Iwan A, Janeczek H, Jarzabek B, Domanski M, Rannou P. Characterization, optical and thermal properties of new azomethines based on heptadecafluoroundecyloxy benzaldehyde. Liquid Crystals. 2009;**36**(8):873-883

[55] Shimoda Y, Ozaki M, Yoshino K. Electric field tuning of a stop band in a reflection spectrum of synthetic opal infiltrated with nematic liquid crystal. Applied Physics Letters. 2001;**79**(22):3627-3629

[56] Gharbi MA, Manet S, Lhermitte J, Brown S, Milette J, Toader V, et al. Reversible nanoparticle cubic lattices in blue phase liquid crystals. ACS Nano. 2016;**10**(3):3410-3415

[57] Ersman PA, Kawahara J, Berggren M. Printed passive matrix addressed electrochromic displays. Organic Electronics. 2013;**14**(12):3371-3378

[58] Chen Y. Flexible active-matrix electronic ink display. Nature. 2003;**423**:136

[59] Hu M-C. Knowledge flows and innovation capability: The patenting trajectory of Taiwan's thin film transistor-liquid crystal display industry. Technological Forecasting and Social Change. 2008;**75**(9):1423-1438

[60] Goodby JW. Twisted and frustrated states of matter. Proceedings of the Royal Society A. 2012;**468**:1521-1542

[61] Lee W-c, DiBenedetto AT. The processing of ternary LCP/LCP/thermoplastic blends. Polymer. 1993;**34**(4):684-690

[62] Shibaev V, Bobrovsky A, Boiko N. Photoactive liquid crystalline polymer systems with light-controllable structure and optical properties. Progress in Polymer Science. 2003;**28**(5):729-836

[63] Yang SY, Kang JH, Jeong SY, Choi KH, Lee S, Shin GJ. Influence of chemical structure on the optical properties of new side-chain polymers containing cholesterol. Liquid Crystals. 2014;**41**(9):1286-1292

[64] Golovoy A, Kozlowski M, Narkis M. Characterization of thermotropic liquid crystalline polyester/polysulfone blends. Polymer Engineering & Science. 1992;**32**(13):854-860

[65] Li C-H, Chang T-C. Studies on thermotropic liquid crystalline polymers—Part II. Synthesis and properties of poly (azomethineether). European Polymer Journal. 1991;**27**(1):35-39

[66] Abdouss M, Afshar Taromi F, Mohammadi N, Ghaffarian SR,

Ranjbarfar B, Jannesari A. Synthesis of main chain liquid crystalline polyesters based on mesogenic monomers with controlled transition temperatures. Iranian Polymer Journal. 2004;**13**(2):121-130

[67] Grell M, Bradley DD, Inbasekaran M, Woo EP. A glass-forming conjugated main-chain liquid crystal polymer for polarized electroluminescence applications. Advanced Materials. 1997;**9**(10):798-802

[68] Xie P, Shen Z, Liu Y, Kong B, Liu C, Zhang R, et al. Synthesis and characterization of a metal chelatebridged quasi-ladder main chain discotic liquid crystal polymer. Liquid Crystals. 2001;**28**(3):477-481

[69] Yu Z, Tu H, Wan X, Chen X, Zhou QF. Synthesis and characterization of mesogen-jacketed liquidcrystal polymers based on 2, 5-bis (4′-alkoxyphenyl) styrene. Journal of Polymer Science Part A: Polymer Chemistry. 2003;**41**(10):1454-1464

[70] Iraqi A, Wataru I. Preparation and properties of 2, 7-linked N-alkyl-9 H-carbazole main-chain polymers. Chemistry of Materials. 2004;**16**(3):442-448

[71] Senthil S, Kannan P. Thermotropic main chain liquid crystalline polyesters containing ferrocene and phosphate units: Synthesis and characterization. Liquid Crystals. 2002;**29**(10):1297-1303

[72] Jeong K-U, Knapp BS, Ge JJ, Jin S, Graham MJ, Xiong H, et al. Structures and phase transformations of asymmetric bent main-chain liquid crystalline polyesters. Macromolecules. 2005;**38**(20):8333-8344

[73] Toh CL, Xu J, Lu X, He C. Synthesis and characterisation of main-chain hydrogen-bonded supramolecular liquid crystalline complexes formed by azo-containing compounds. Liquid Crystals. 2008;**35**(3):241-251

[74] Rim M, Kang D-G, Jung D, Lim S-I, Lee KM, Godman NP, et al. Remote-controllable and encryptable smart glasses: A photoresponsive azobenzene molecular commander determines the molecular alignments of liquid crystal soldiers. Nanoscale. 2022;**14**(23):8271-8280

[75] Bagheri M, Rad RZ. Synthesis and characterization of thermotropic liquid crystalline polyesters with biphenyl unit in the main chain. Reactive and Functional Polymers. 2008;**68**(2):613-622

[76] Walba DM, Yang H, Keller P, Zhu C, Shao R, Coleman DA, et al. A Mainchain de Vries Smectic liquid crystal polymer prepared by Hoveyda–Grubbs catalyst initiated acyclic diene metathesis polymerization. Macromolecular Rapid Communications. 2009;**30**(22):1894-1899

[77] Yang H, Buguin A, Taulemesse J-M, Kaneko K, Méry S, Bergeret A, et al. Micron-sized main-chain liquid crystalline elastomer actuators with ultralarge amplitude contractions. Journal of the American Chemical Society. 2009;**131**(41):15000-15004

[78] Ohkawa S, Ohta R, Kawabata K, Goto H. Polymerization in liquid crystal medium: Preparation of polythiophene derivatives bearing a bulky pyrimidine substituent. Polymers. 2010;**2**(4):393-406

[79] Amela-Cortés M, Bruce DW, Evans KE, Smith CW. Unsymmetric main-chain liquid crystal elastomers with tunable phase behaviour: Elastic response. Journal of Materials Chemistry. 2011;**21**(23):8436-8442

[80] Gimeno N, Sánchez-Ferrer A, Sebastián N, Mezzenga R, Ros MB. Bent-core based main-chain polymers showing the dark conglomerate liquid crystal phase. Macromolecules. 2011;**44**(24):9586-9594

[81] Kinoshita M, Takamura K, Kawamoto M, Shishido A, Shiono T, Ikeda T. Synthesis and emission behavior of liquid-crystalline Main-chain polyesters containing Carbazole and Oxadiazole moieties. Molecular Crystals and Liquid Crystals. 2012;**563**(1):92-100

[82] Al-Muaikel NS, Aly KI. Liquid crystalline polymers xiii main chain thermotropic copoly (arylideneether)s containing 4-teriary butylcyclohexanone moiety linked with polymethylene spacers. Open Journal of Organic Polymer Materials. 2013;**3**(1):19

[83] Fang L, Zhang H, Li Z, Zhang Y, Zhang Y, Zhang H. Synthesis of reactive azobenzene main-chain liquid crystalline polymers via Michael addition polymerization and photomechanical effects of their supramolecular hydrogenbonded fibers. Macromolecules. 2013;**46**(19):7650-7660

[84] Samui AB, Pandey S, Mishra SP. Main chain photoresponsive liquid crystalline polymer synthesized through hydrosilylation. RSC Advances. 2015;**5**(84):68351-68355

[85] Saed MO. Structure-Property Relationships in Thiol-Acrylate Based Main-Chain Liquid-Crystalline Elastomers. Denver (USA): University of Colorado; 2017

[86] Jo B-W, Jin J-I, Lenz R. Liquid crystal polymers—VI.: Synthesis and properties of main chain thermotropic polyesters with disiloxane spacers. European Polymer Journal. 1982;**18**(3):233-239

[87] Yang C-A, Wang Q, Xie H, Zhong G, Zhang H. Synthesis and characterisation of polymethacrylates containing para-, meta-and ortho-monosubstituted azobenzene moieties in the side chain. Liquid Crystals. 2010;**37**(10):1339-1346

[88] Crivello JV, Deptolla M, Ringsdorf H. The synthesis and characterization of side-chain liquid crystal polymers based on polystyrene and poly-α-methystyrene. Liquid Crystals. 1988;**3**(2):235-247

[89] Andreozzi L, Faetti M, Giordano M, Palazzuoli D. Physical aging in sidechain liquid crystal polymers: A DSC investigation of the enthalpy relaxation. Kuala Lumpur (Malaysia). Macromolecules. 2002;**35**(24):9049-9056

[90] Guittard F, de Givenchy ET, Geribaldi S, Cambon A. Highly fluorinated thermotropic liquid crystals: An update. Journal of Fluorine Chemistry. 1999;**100**(1):85-96

[91] Mohamad SN. Design of Polymerizable Liquid Crystal Monomers for New Side Chain Liquid Crystalline Polymers (SCLCPs) and their Properties. Kuala Lumpur (Malaysia): University of Malaya; 2014

[92] Finkelmann H, Happ M, Portugal M, Ringsdorf H. Liquid crystalline polymers with biphenyl-moieties as mesogenic group. Macromolecular Chemistry and Physics. 1978;**179**(10):2541-2544

[93] Kreuder W, Ringsdorf H. Liquid crystalline polymers with disc-like mesogens. Macromolecular Rapid Communications. 1983;**4**(12):807-815

[94] Percec V, Barboiu B. "Living" radical polymerization of styrene initiated by arenesulfonyl chlorides and CuI (bpy) nCl. Macromolecules. 1995;**28**(23):7970-7972

[95] Barner-Kowollik C, Davis TP, Heuts J, Stenzel MH, Vana P, Whittaker M. Rafting down under: Tales of missing radicals, fancy architectures, and mysterious holes. Journal of Polymer Science Part A: Polymer Chemistry. 2003;**41**(3):365-375

[96] Ganicz T, Stańczyk W. Side-chain liquid crystal polymers (SCLCP): Methods and materials. An overview. Materials. 2009;**2**(1):95-128

[97] Ujiie S, Kato T. Design and synthesis of side-chain liquid crystal polymers. In: Goodby JW, Collings PJ, Kato T, Tschierske C, Gleeson HF, Raynes P, editors. Handbook of Liquid Crystals. 1st ed. Weinheim (Germany): WileyVCH; 2014

[98] Komiya Z, Schrock RR. Synthesis of side chain liquid crystal polymers by living ring-opening metathesis polymerization. 5. Influence of mesogenic group and interconnecting group on the thermotropic behavior of the resulting polymers. Macromolecules. 1993;**26**(6):1393-1401

[99] Cowling SJ, Toyne KJ, Goodby JW. Side-chain liquid crystal polymers derived from oxetane monomers containing 2-or 3-fluorophenyl moieties in the core of the mesogen. Journal of Materials Chemistry. 2001;**11**(6):1590-1599

[100] Sato M, Mizoi M. Relationships between structures and liquid crystalline and optical properties in combined-type copolyisophthalates having nitro-and methoxyazobenzene units and aliphatic spacers in the side chain. Polymer Journal. 2004;**36**(8):607-616

[101] Han X, Shanks RA, Pavel D. The synthesis and thermal properties of polyepichlorohydrin side-chain liquid crystal polymers. European Polymer Journal. 2005;**41**(5):984-991

[102] Han X, Chai L, Shanks R, Pavel D. Monte Carlo simulations of properties of side-chain liquid-crystal polymers. Polymer International. 2006;**55**(11):1323-1329

[103] Verploegen E, Zhang T, Jung YS, Ross C, Hammond PT. Controlling the morphology of side chain liquid crystalline block copolymer thin films through variations in liquid crystalline content. Nano Letters. 2008;**8**(10):3434-3440

[104] Kang H, Lee J-C, Kang D. The effect of phenoxymethyl side groups on the liquid crystal alignment behavior of polystyrene derivatives. Macromolecular Research. 2009;**17**(7):506-515

[105] Aya S, Haba O, Yonetake K, Araoka F. Anchoring and molecular conformation of liquid crystalline dendrimer. Journal of Molecular Liquids. 2021;**321**:114379

[106] Wang F, Zhu G, Gong X, Geng P, Gao Z, Wang Y. The impact of the flexibility of photosensitive polyimide backbone on the properties of liquidcrystal alignment under non-polarised ultraviolet light. Liquid Crystals. 2021;**48**(8):1111-1119

[107] Mohammed IA, Sankar G, Khairuddean M, Mohamad AB. Synthesis and liquid crystalline properties of new diols containing azomethine groups. Molecules. 2010;**15**(5):3260-3269

[108] Park JS, Kim YH, Song M, Kim CH, Karim M, Lee JW, et al. Synthesis and photovoltaic properties of side-chain liquid-crystal click polymers for dyesensitized solar-cells application. Macromolecular Chemistry and Physics. 2010;**211**(23):2464-2473

[109] Petr M, Hammond PT. Room temperature rapid photoresponsive

azobenzene side chain liquid crystal polymer. Macromolecules. 2011;**44**(22):8880-8885

[110] Smokal V, Krupka O, Sahraoui B, Kolendo A. Side-chain Azobenzene polymers: Synthesis and photochemical properties. Molecular Crystals and Liquid Crystals. 2011;**536**(1):115/[347]-121/[353]

[111] Hofmann AM, Wipf R, Stühn B, Frey H. Mesogen-initiated linear polyglycerol isomers: The ordering effect of a single cholesterol unit on "sticky" isotropic chains. Macromolecules. 2011;**44**(17):6767-6775

[112] Kondo M, Dozono Y, Goto K, Kawatsuki N. Synthesis of side-chain liquid-crystalline Cinnamide polymers based on post polymer reaction. Molecular Crystals and Liquid Crystals. 2012;**563**(1):121-130

[113] Nepal S, Das B, Das MK, Sarkar MD, Strójwąs K, Dmochowska E, et al. High tilted antiferroelectric liquid crystals: Polymer-based approach for phase stabilisation and device development. Journal of Molecular Liquids. 2023;**375**:121297

[114] Ban J, Chen S, Zhang H. Synthesis and phase behavior of mesogenjacketed liquid crystalline polymer with triphenylene discotic liquid crystal mesogen unit in side chains. RSC Advances. 2014;**4**(97):54158-54167

[115] Wu B, Mu B, Wang S, Duan J, Fang J, Cheng R, et al. Triphenylenebased side chain liquid crystalline block copolymers containing a PEG block: Controlled synthesis, microphase structures evolution and their interplay with discotic mesogenic orders. Macromolecules. 2013;**46**(8):2916-2929

[116] Xie J, Zhu G, Tang Y, Zhang Z. Synthesis and properties of reactive liquid crystal monomers and side-chain liquid crystalline polymers. Liquid Crystals. 2013;**40**(4):546-554

[117] Yang C-a, Xie H, Chen S, Zhang H. Synthesis and phase structures of combined main-chain/side-chain liquid crystalline polymers with different-number azobenzene moiety in the side-chain based on mesogenjacketed liquid crystalline polymers. Liquid Crystals. 2013;**40**(11):1487-1502

[118] Gimeno N, Pintre I, Martínez-Abadía M, Serrano JL, Ros MB. Bent-core liquid crystal phases promoted by azo-containing molecules: From monomers to side-chain polymers. RSC Advances. 2014;**4**(38):19694-19702

[119] Hirai T, Osumi S, Ogawa H, Hayakawa T, Takahara A, Tanaka K. Precise synthesis and surface wettability of a polymer with liquid crystalline side chains. Macromolecules. 2014;**47**(15):4901-4907

[120] Hu J-S, Zhang W-C, Su D, Guo Z-H. New side chain ferroelectric liquid crystalline polymers based on (2 S, 3 S)-2-chloro-3-methylpentanoic acid: Synthesis and phase behaviour. Liquid Crystals. 2014;**41**(11):1526-1536

[121] Kandasamy P, Keerthiga R, Vijayalakshmi S, Kaliyappan T. Synthesis and liquid crystal properties of supramolecular side-chain liquidcrystalline polymers containing poly (acrylic acid) intermolecular hydrogen bonds. Molecular Crystals and Liquid Crystals. 2015;**606**(1):1-11

[122] Yao W, Gao Y, Yuan X, He B, Yu H, Zhang L, et al. Synthesis and self-assembly behaviours of sidechain smectic thiol–ene polymers based on the polysiloxane backbone. Journal of Materials Chemistry C. 2016;**4**(7):1425-1440

[123] Zhang H, Miao Z, Shen W. Development of polymer-dispersed liquid crystals: From mode innovation to applications. Composites Part A: Applied Science and Manufacturing. 2022;**163**:107234

[124] Perju E, Paslaru E, Marin L. Polymer-dispersed liquid crystal composites for bio-applications: Thermotropic, surface and optical properties. Liquid Crystals. 2015;**42**(3):370-382

[125] Mach P, Rodriguez S, Nortrup R, Wiltzius P, Rogers JA. Monolithically integrated, flexible display of polymerdispersed liquid crystal driven by rubber-stamped organic thin-film transistors. Applied Physics Letters. 2001;**78**(23):3592-3594

[126] Cupelli D, Nicoletta FP, Manfredi S, Vivacqua M, Formoso P, De Filpo G, et al. Self-adjusting smart windows based on polymer-dispersed liquid crystals. Solar Energy Materials and Solar Cells. 2009;**93**(11):2008-2012

[127] Lee JW, Kim JK, Ahmad F, Jamil M, Jeon YJ. Properties of thiol–vinyl PDLC films without additional photoinitiator. Liquid Crystals. 2014;**41**(8):1109-1115

[128] Malik M, Bhatia P, Deshmukh R. Effect of nematic liquid crystals on optical properties of solvent induced phase separated PDLC composite films. Indian Journal of Science and Technology. 2012;**5**(10):3440-3452

[129] Ren H, Lin Y-H, Fan Y-H, Wu S-T. Polarization-independent phase modulation using a polymer-dispersed liquid crystal. Applied Physics Letters. 2005;**86**(14):141110

[130] Xu J, Li H, Guo L, Zhang Z, Zou C, Wang Q, et al. Effect of curing temperature on the properties of electrically controlled dimming film with wide working temperature range. Crystals. 2022;**12**(11):1551

[131] Marin L, Ailincai D, Paslaru E. Monodisperse PDLC composites generated by use of polyvinyl alcohol boric acid as matrix. RSC Advances. 2014;**4**(72):38397-38404

[132] Jeong D-H, Jung Ch-W, Jung K-S, Hong C-K. Dynamic characteristics of the PDLC-based electrooptic modulator for TFT-LCD inspection. In: Proc. SPIE 4902, Optomechatronic Systems III. 18 Oct 2002. DOI: 10.1117/12.467530

[133] Amimori I, Priezjev NV, Pelcovits RA, Crawford GP. Optomechanical properties of stretched polymer dispersed liquid crystal films for scattering polarizer applications. Journal of Applied Physics. 2003;**93**(6):3248-3252

[134] Masutani A, Roberts T, Schüller B, Yasuda A, Sakaigawa A, Cross G, et al. A novel polarizer-free dye-doped polymerdispersed liquid crystal for reflective TFT displays. Journal of the Society for Information Display. 2004;**12**(3):301-307

[135] Nicoletta F, De Filpo G, Lanzo J, Chidichimo G. A method to produce reverse-mode polymer-dispersed liquidcrystal shutters. Applied Physics Letters. 1999;**74**(26):3945-3947

[136] Liu Y, Sun X, Dai H, Liu J, Xu K. Effect of surfactant on the electro-optical properties of holographic polymer dispersed liquid crystal Bragg gratings. Optical Materials. 2005;**27**(8):1451-1455

[137] Shim S, Woo J, Kim K, Kim B. Effects of surfactant and molecular weight of polyol on grating formation and switching of holographic PDLC.

Polymers for Advanced Technologies. 2008;**19**(11):1550-1557

[138] Hinojosa A, Sharma SC. Effects of gold nanoparticles on electro-optical properties of a polymer-dispersed liquid crystal. Applied Physics Letters. 2010;**97**(8):081114

[139] Tsai T-Y, Lee C-Y, Lee C-J, Chang W-C, Lee W, Chen P-C. Preparation and characterization of polymer-dispersed liquid crystal (PDLC) with different cation exchange capacity (CEC) clays. Journal of Physics and Chemistry of Solids. 2010;**71**(4):595-599

[140] Liu G, Xia D, Yang W, Huang Z. The surface rubbing effect on morphologies of LC droplets and electro-optic properties of flexible PDLC films. Science in China Series B: Chemistry. 2009;**52**(12):2329-2335

[141] Macchione M, Cupelli D, Filpo GD, Nicoletta FP, Chidichimo G. Rough surfaces for orientation control in reverse mode polymer dispersed liquid crystal films. Liquid Crystals. 2000;**27**(7):917-920

[142] Lee HG, Lee CS, Lee SG, Heo YJ, Lim YJ, Lee SH. Guest-host liquid crystal devices utilizing dichroic dye and its application. SID Symposium Digest of Technical Papers. 2019;**50**(1):1652-1655

[143] Chen W-T, Ji S-C, Chen S-H, Huang C-Y, Lo KY. Competitive dye adsorption-desorption on the isotropic surface at the early stage of the photoexcitation of azo dye-doped liquid crystals. Crystals. 2020;**10**(9):802-812

[144] Tian L-L, Chu F, Duan W, Li R, Gu X-Q, Li L, et al. Beam steering device based on blue phase liquid crystal. Optics Communications. 2021;**481**:126525

[145] Brás AR, Casimiro T, Caldeira J, Aguiar-Ricardo A. Solubility of the nematic liquid crystal E7 in supercritical carbon dioxide. Journal of Chemical & Engineering Data. 2005;**50**(6):1857-1860

[146] Ren H, Fan Y-H, Lin Y-H, Wu S-T. Tunable-focus microlens arrays using nanosized polymer-dispersed liquid crystal droplets. Optics Communications. 2005;**247**(1):101-106

[147] Kim C, Son I, Kim JH, Yoo JY, Lee B, Lee JH. New hydrogen-bonded comblike polyimides with mesogenic side chains for stable orientation and fast switching of liquid crystals. Liquid Crystals. 2020;**47**(5):761-767

[148] Sumana G, Raina K. Electro-optic properties of aligned polysiloxane dispersed ferroelectric liquid crystal composite thin films. Current Applied Physics. 2005;**5**(3):277-284

[149] Choi J-W, Hee Kim S, Jin Shin G, Kim J-W, Hyun Kim D, Hyup LJ. Hierarchically self-constructed alignment layer of comb-shaped amphiphilic copolymers for spontaneous and stable vertical orientation of liquid crystals. Journal of Molecular Liquids. 2023;**373**:121247

[150] Senyurt AF, Warren G, Whitehead JB, Hoyle CE. Matrix physical structure effect on the electrooptic characteristics of thiol–ene based H-PDLC films. Polymer. 2006;**47**(8):2741-2749

[151] Kumar P, Raina K. Morphological and electro-optical responses of dichroic polymer dispersed liquid crystal films. Current Applied Physics. 2007;**7**(6):636-642

[152] Petrov A, Marinov Y, Delia S, Marino S, Versace C, Scaramuzza N. Dielectric and flexoelectric oscillations in PDLC studied by flexoelectric spectroscopy and laser light diffraction.

Journal of Optoelectronics and Advanced Materials. 2007;**9**(2):420

[153] He J, Yan B, Yu B, Wang S, Zeng Y, Wang Y. The effect of molecular weight of polymer matrix on properties of polymer-dispersed liquid crystals. European Polymer Journal. 2007;**43**(6):2745-2749

[154] Wang P-C, MacDiarmid AG. Integration of polymer-dispersed liquid crystal composites with conducting polymer thin films toward the fabrication of flexible display devices. Displays. 2007;**28**(3):101-104

[155] Jin F-L, Park S-J. Effect of fluorine group on electro-optical properties of PDLC films prepared with watersoluble polymers. Bulletin of the Korean Chemical Society. 2008;**29**(12):2521-2524

[156] Ding X, Cao M, Liu H, Cao H, Li W, Yang H. A study of electro-optical properties of PDLC films prepared by dual UV and heat curing. Liquid Crystals. 2008;**35**(5):587-595

[157] Li W, Cao Y, Cao H, Kashima M, Kong L, Yang H. Effects of the structures of polymerizable monomers on the electro-optical properties of UV cured polymer dispersed liquid crystal films. Journal of Polymer Science Part B: Polymer Physics. 2008;**46**(13):1369-1375

[158] Koo JJ, No Y-S, Jeon C-W, Kim JH. Improvement of electro-optic properties in PDLC device by using new crosslinker for the control of the contrast ratio, response time and driving voltage. Molecular Crystals and Liquid Crystals. 2008;**491**(1):58-66

[159] Yan B, He J, Bao R, Bai X, Wang S, Zeng Y, et al. Modification of electrooptical properties of polymer dispersed liquid crystal films by iniferter polymerization. European Polymer Journal. 2008;**44**(3):952-958

[160] Lallana P, Vázquez C, Vinouze B, Heggarty K, Montero D. Multiplexer and variable optical attenuator based on PDLC for polymer optical fiber networks. Molecular Crystals and Liquid Crystals. 2009;**502**(1):130-142

[161] Bacchiocchi C, Miglioli I, Arcioni A, Vecchi I, Rai K, Fontecchio A, et al. Order and dynamics inside H-PDLC nanodroplets: An ESR spin probe study. The Journal of Physical Chemistry B. 2009;**113**(16):5391-5402

[162] Nataj NH, Mohajerani E, Jashnsaz H, Jannesari A. Holographic polymer dispersed liquid crystal enhanced by introducing urethane trimethacrylate. Applied Optics. 2012;**51**(6):697-703

[163] Natarajan LV, Shepherd CK, Brandelik DM, Sutherland RL, Chandra S, Tondiglia VP, et al. Switchable holographic polymer-dispersed liquid crystal reflection gratings based on thiolene photopolymerization. Chemistry of Materials. 2003;**15**(12):2477-2484

[164] Chidichimo G, De Filpo G, Manfredi S, Mormile S, Tortora L, Gallucci C, et al. High contrast reverse mode PDLC films: A morphologic and electro-optical analysis. Molecular Crystals and Liquid Crystals. 2009;**500**(1):10-22

[165] Kashima M, Cao H, Meng Q, Liu H, Wang D, Li F, et al. The influence of crosslinking agents on the morphology and electro-optical performances of PDLC films. Journal of Applied Polymer Science. 2010;**117**(6):3434-3440

[166] Hsiao V, Chang W-T. Optically switchable, polarization-independent

holographic polymer dispersed liquid crystal (H-PDLC) gratings. Applied Physics B. 2010;**100**(3):539-546

[167] Lin Y-H, Tsou Y-S. A polarization independent liquid crystal phase modulation adopting surface pinning effect of polymer dispersed liquid crystals. Journal of Applied Physics. 2011;**110**(11):114516

[168] Yu H, Dong C, Zhou W, Kobayashi T, Yang H. Wrinkled liquidcrystalline microparticle-enhanced photoresponse of PDLC-like films by coupling with mechanical stretching. Small. 2011;**7**(21):3039-3045

[169] Jung JE, Lee GH, Jang JE, Hwang KY, Ahmad F, Muhammad J, et al. Optical enhancement of dye-doped PDLC by additional dye-LC layer coating. Optical Materials. 2011;**34**(1):256-260

[170] Su Y-C, Chu C-C, Chang W-T, Hsiao V. Characterization of optically switchable holographic polymerdispersed liquid crystal transmission gratings. Optical Materials. 2011;**34**(1):251-255

[171] Parab S, Malik M, Deshmukh R. Dielectric relaxation and electro-optical switching behavior of nematic liquid crystal dispersed in poly (methyl methacrylate). Journal of Non-Crystalline Solids. 2012;**358**(18):2713-2722

[172] Wang J, Shi S, Han X, Zhou S, Zhang B. The effect of nanographites dispersed in Undec-10-enic acid on optical property and morphology of PDLC films. Colloid and Polymer Science. 2012;**290**(16):1659-1664

[173] Ahmad F, Luqman M, Jamil M. Advances in the metal nanoparticles (MNPs) doped liquid crystals and polymer dispersed liquid crystal (PDLC) composites and their applications - a review. Molecular Crystals and Liquid Crystals. 2021;**731**(1):1-33

[174] Shivaraja SJ, Gupta RK, Kumar S, Manjuladevi V. Enhanced electro-optical response of nematic liquid crystal doped with functionalised silver nanoparticles in twisted nematic configuration. Liquid Crystals. 2020;**47**(11):1678-1690

[175] Kumar A, Singh G, Joshi T, Biradar AM. Electro-optical and dielectric characteristics of ferroelectric liquid crystal dispersed with palladium nanoparticles. Journal of Molecular Liquids. 2020;**315**:113776

[176] Meng X, Li J, Lin Y, Liu X, Zhao J, Li G, et al. Effect of the magnetic nanoparticle concentration on the electro-optical properties of the polymer-dispersed liquid crystals. Liquid Crystals. 2022;**49**(12):1612-1622

[177] Ahmad F, Jeon AR, Jeon YJ, Jamil M. A novel technique of fabrication of nanoparticle acrylate doped polymer dispersed liquid crystal (PDLC) film. Journal of Dispersion Science and Technology. 2022;**43**(10):1506-1511

[178] Zito G, Pissadakis S. Holographic polymer-dispersed liquid crystal Bragg grating integrated inside a solid core photonic crystal fiber. Optics Letters. 2013;**38**(17):3253-3256

[179] Parab S, Malik M, Bhatia P, Deshmukh R. Investigation of liquid crystal dispersion and dielectric relaxation behavior in polymer dispersed liquid crystal composite films. Journal of Molecular Liquids. 2014;**199**:287-293

[180] Yu L, Cheng Z, Dong Z, Zhang Y, Yu H. Photomechanical response of polymer-dispersed liquid crystals/ graphene oxide nanocomposites.

Journal of Materials Chemistry C. 2014;**2**(40):8501-8506

[181] Eun Jung J, Lee GH, Eun Jang J, Hwang KY, Ahmad F, Jamil M, et al. Optical property enhancement of dyePDLC using active reflector structure. Journal of Applied Polymer Science. 2012;**124**(1):873-877

[182] Deshmukh R, Malik M, Parab S. Dichroic dye induced nonlinearity in polymer dispersed liquid crystal materials for display devices. Advanced Materials Research. 2012;**584**(1):79-83

[183] Lu Y, Wei J, Shi Y, Jin O, Guo J. Effects of fabrication condition on the network morphology and electro-optical characteristics of polymer-dispersed bistable smectic a liquid crystal device. Liquid Crystals. 2013;**40**(5):581-588

[184] Ma Z, Gao Y, Cao H. The effect of chemically modified multi-walled carbon nanotubes on the electrooptical properties of a twisted Nematic liquid crystal display mode. Crystals. 2022;**12**(10):1482

[185] Pagidi S, Manda R, Bhattacharyya SS, Cho KJ, Kim TH, Lim YJ, et al. Superior electro-optics of nano-phase encapsulated liquid crystals utilizing functionalized carbon nanotubes. Composites Part B: Engineering. 2019;**164**:675-682

[186] Yadav G, Agrahari K, Manohar R. Multiwall carbon nanotube-nematic liquid crystal composite system: Preparation and characterization. Journal of Dispersion Science and Technology. 2021;**42**(5):707-714

[187] Xie A-Q, Cui T, Cheng R, Wu X, Guo J, Lu X, et al. Robust nanofiber films prepared by electro-microfluidic spinning for flexible highly stable quantum-dot displays. Advanced Electronic Materials. 2021;**7**(1):2000626

[188] Singh G, Fisch M, Kumar S. Tunable polarised fluorescence of quantum dot doped nematic liquid crystals. Liquid Crystals. 2016;**44**:1-9

[189] Zhang S, Li C, Wang Q, Zhou L, Saeed MH, Wang X, et al. Fluorescence enhancement and encapsulation of quantum dots via a novel crosslinked vinyl-ether liquid crystals/ polymer composite film. Polymer. 2020;**207**:122834

[190] Kim Y, Jung D, Jeong S, Kim K, Choi W, Seo Y. Optical properties and optimized conditions for polymer dispersed liquid crystal containing UV curable polymer and nematic liquid crystal. Current Applied Physics. 2015;**15**(3):292-297

[191] Song P, Gao Y, Wang F, Zhang L, Xie H, Yang Z, et al. Studies on the electro-optical and the light-scattering properties of PDLC films with the size gradient of the LC droplets. Liquid Crystals. 2015;**42**(3):390-396

[192] Khaligh HH, Liew K, Han Y, Abukhdeir NM, Goldthorpe IA. Silver nanowire transparent electrodes for liquid crystal-based smart windows. Solar Energy Materials and Solar Cells. 2015;**132**:337-341

[193] Ma H, Zhou L, Han C, Zhang C, Zhang L. The fabrication of novel optical diffusers based on UV-cured polymer dispersed liquid crystals. Liquid Crystals. 2019;**46**(1):138-144

[194] Kushawaha R, Sinha A. A facile fabrication of ultrahigh haze optical diffuser using nematic liquid crystal encapsulated by cellulose acetate biopolymer. Journal of Physics D: Applied Physics. 2022;**55**(28):285303

[195] He Z, Yin K, Hsiang E-L, Wu S-T. Volumetric light-shaping

polymerdispersed liquid crystal films for mini-LED backlights. Liquid Crystals. 2020;**47**(10):1458-1463

[196] Parsley M. The Hallcrest Handbook of Thermochromic Liquid Crystal Technology. Glenview: Hallcrest; 1991

[197] Najjar R, Bigdeli E, Asadpour-Zeynali K, Zaker-Hamidi M-S. Liquid crystalline nanomaterials extracted from egg yolk: Encapsulation and characterization of their electro-optical activity as potential materials for flexible LCD displays. Journal of Electronic Materials. 2018;**47**(12):7143-7150

[198] Moghbeli M, Abedi V, Dekamin M. Microencapsulation of Ethion by interfacial polymerization utilizing potassium Phthalimide-N-oxyl (PPINO) as a promoter. Iranian Journal of Chemical Engineering. 2011;**8**(4):35

[199] Najjar R, Bigdeli E. Synthesis of novel core-shells of PMMA with coumarin based liquid crystalline side chains and PMMA shell as electro-optical materials. European Polymer Journal. 2018;**104**:136-146

[200] Kim M, Park KJ, Seok S, Ok JM, Jung H-T, Choe J, et al. Fabrication of microcapsules for dye-doped polymerdispersed liquid crystal-based smart windows. ACS Applied Materials & Interfaces. 2015;**7**(32):17904-17909

[201] Guo J, Zhang J, Zhang Q, Jiang N, Wei J. Fabrication of cholesteric liquid crystal microcapsulates by interfacial polymerization and potential as photonic materials. RSC Advances. 2013;**3**(44):21620-21627